Theoretical Aspects and Computer Modeling of the Molecular Solid State

Editorial Board

Theoretical Aspects and Computer Modeling of the Molecular Solid State

Edited by
Angelo Gavezzotti
University of Milan, Italy

The Molecular Solid State Volume 1

JOHN WILEY & SONS
Chichester · New York · Weinheim · Brisbane · Singapore · Toronto

Baffins Lane, Chichester,
West Sussex PO19 1UD, England

National 01243 779777
International (+44) 1243 779777
e-mail (for orders and customer service enquiries): cs-book@wiley.co.uk
Visit our Home Page on http://www.wiley.co.uk
or http://www.wiley.com

Other Wiley Editorial Offices

John Wiley & Sons, Inc., 605 Third Avenue,
New York, NY 10158-0012, USA

VCH Verlagsgesellschaft mbH, Pappelallee 3,
D-69469 Weinheim, Germany

Jacaranda Wiley Ltd, 33 Park Road, Milton,
Queensland 4064, Australia

John Wiley & Sons (Asia) Pte Ltd, 2 Clementi Loop #02-01,
Jin Xing Distripark, Singapore 129809

John Wiley & Sons (Canada) Ltd, 22 Worcester Road,
Rexdale, Ontario M9W 1L1, Canada

Library of Congress Cataloging-in-Publication Data

The molecular solid state.
p. cm.
Includes bibliographical references and indexes.
Contents: v. 1. Theoretical aspects and computer modeling/ edited by Angelo Gavezzotti.
ISBN 0-471-96187-6 (v. 1)
1. Molecular crystals. I. Gavezzotti, Angelo.
QD921.M683 1997
541.2′2–dc20 96-30338
CIP

British Library Cataloguing in Publication Data

A catalogue record for this book is available from the British Library

ISBN 0-471-96187-6

Typeset in 10/12pt Times by Keytec Typesetting Ltd, Bridport, Dorset
Printed and bound in Great Britain by Biddles, Guildford, Surrey
This book is printed on acid-free paper responsibly manufactured from sustainable forestation, for which at least two trees are planted for each one used for paper production.

Contents

Contributors vii

Preface ix

1 **Crystal Symmetry and Molecular Recognition** 1
A. Gavezzotti

2 **Intermolecular Forces – from the Molecular Charge Distribution to the Molecular Packing** 31
S. L. Price

3 **Energetic Aspects of Crystal Packing: Experiment and Computer Simulations** 61
A. Gavezzotti and G. Filippini

4 **Energy Minimization and Molecular Dynamics Calculations for Molecular Crystals** 99
B. P. van Eijck, L. M. J. Kroon-Batenburg and J. Kroon

5 **Nucleation and Phase Transitions in Molecular Clusters: Molecular Dynamics Simulation and Experiment** 147
L. S. Bartell

6 ***Ab Initio* Prediction of Possible Molecular Crystal Structures** 185
R. J. Gdanitz

7 **The Crystal Habit of Molecular Materials: A Structural Perspective** 203
G. Clydesdale, K. J. Roberts and E. M. Walker

Index 233

Contributors

Lawrence S. Bartell, Department of Chemistry, University of Michigan, Ann Arbor, Michigan 48109, USA

Graham Clydesdale, Centre for Molecular and Interface Engineering, Department of Mechanical and Chemical Engineering, Heriot-Watt University, Riccarton, Edinburgh, EH14 4AS

Giuseppe Filippini, Centro CNR per lo Studio delle Relazioni tra Struttura e Reattività Chimica, Università di Milano, Milano, Italy

Angelo Gavezzotti, Dipartimento di Chimica Strutturale e Stereochimica Inorganica, Università di Milano, via Venezian 21, 20133 Milano, Italy

Robert J. Gdanitz, Fachbereich Physik der Gesamthochschule Kassel, Heinrich-Plett-Strasse 40, D-34132 Kassel, Germany

Jan Kroon, Department of Crystal and Structural Chemistry, Bijvoet Center for Biomolecular Research, Utrecht University, Padualaan 8, 3584 CH Utrecht, The Netherlands

Loes M. J. Kroon-Batenburg, Department of Crystal and Structural Chemistry, Bijvoet Center for Biomolecular Research, Utrecht University, Padualaan 8, 3584 CH Utrecht, The Netherlands

S. L. Price, Department of Chemistry, University College London, 20 Gordon Street, London, WC1H 0AJ

Kevin J. Roberts, Centre for Molecular and Interface Engineering, Department of Mechanical and Chemical Engineering, Heriot-Watt University, Riccarton, Edinburgh, EH14 4AS

Bouke P. van Eijck, Department of Crystal and Structural Chemistry, Bijvoet Center for Biomolecular Research, Utrecht University, Padualaan 8, 3584 CH Utrecht, The Netherlands

Elaine M. Walker, Centre for Molecular and Interface Engineering, Department of Mechanical and Chemical Engineering, Heriot-Watt University, Riccarton, Edinburgh, EH14 4AS

Preface

New books cannot always please, but the appearance of this series of titles on the molecular solid state will be welcomed by many scientists. The description, molecular solid state, is taken to apply to solids where the forces between neighboring particles are much weaker than the intraparticle forces; thus we distinguish molecular solids from ionic, covalent and metallic solids, where interparticle forces are strong. Since we hope to discuss liquids and even clusters, perhaps a more accurate title for our series would have been: Molecular Condensed States. However, there is no question that our understanding of crystalline solids is much better developed than that of less ordered systems, so we have remained with the 'solid state'. In any case, the emphasis will be on the weak intermolecular interactions that are responsible for the very existence of the organic world.

Quantum mechanics may be the only fundamental theory applicable at this level, but properties of condensed matter have their own laws that can seldom be derived, in a strict reductionist sense, from the fundamental theory. New experimental facts continue to surprise us, and we have a wealth of non-rigorous semi-empirical models with which to discuss them.

As far as crystalline solids are concerned, we have an enormous database of known crystal structures, which is presently growing at the rate of something in excess of 10 000 structures a year. With the advent of various kinds of area detectors and still more efficient computers, this mass production of crystallographic data is going to increase still further. The main instrument for studying the systematics of crystal structures is the Cambridge Structural Database (CSD), compiled at the Cambridge Crystallographic Data Centre (CCDC), founded by Dr Olga Kennard more than thirty years ago. While the main emphasis used to be on the molecules themselves (interatomic distances, torsion angles, configuration, etc.), the focus of interest is moving increasingly towards the systematics of intermolecular interactions. It is a complex subject; one has to look at many crystal packings to identify the recurring patterns and the exceptions from these and the reasons for these exceptions.

Besides the available experimental crystal structures, we have a great variety of models to account for them, or at least for recurring salient features. These models range from elaborate ones comprising systems of point charges, dipoles, quadrupoles, etc., derived from *ab initio* quantum mechanical calculations and including also second-order polarization and dispersion at the one extreme, to outspokenly simple ones, designed to reproduce a limited set of data, at the other. One of the arts is to decide what level of complexity is called for to handle a given type of problem. In general, the simpler the problem, the more complicated the underlying model. Thus, for a prediction (or postdiction) of the crystal structure of molecular chlorine we need to employ a much more elaborate model than the simple pair potential model that we might use in a discussion at the atomic level of some detail of a protein structure. But at the bottom of all models of crystal structure lies the realization that the long-range periodic order of a crystal is a consequence of short-range interactions between neighboring molecules. Although much progress has been made (as described in this volume), we are still far from the goal of predicting crystal structures in general. Nevertheless, typical patterns (the subject matter of 'molecular recognition', a thriving branch of supramolecular chemistry), such as those involved in hydrogen bonding, recur and are actively being sought and used for the design of solids with desired properties.

The present volume, edited by Professor Angelo Gavezzotti, covering theoretical aspects and computer modeling, deals both with the possibilities and with the limitations of these approaches, and seems to me to be an excellent send-off to the series.

J. D. Dunitz

Chapter 1

Crystal Symmetry and Molecular Recognition

A. GAVEZZOTTI
Dipartimento di Chimica Strutturale e Stereochimica Inorganica, Università di Milano

1. CRYSTAL SYMMETRY

1.1 Introduction

Much to the astonishment of physicists experimenting with X-ray diffraction in the first decades of this century, solid matter turned out to be mostly crystalline. This simple discovery not only led the way to modern structural chemistry, but also somewhat changed the current views on the organization of matter, whose atomic and molecular constituents unexpectedly showed a high propensity for symmetrical ordering, rather than random orientation, when in close proximity to each other. Once again, symmetry proved to play a key role in our understanding of the physical world. In fact, no theoretical aspects of crystals can be understood, and no computer modeling of crystals can be performed, without a thorough appreciation of their symmetry, meaning here the regular, infinitely self-repeating pattern of organization of chemical entities in space. To the theoretical chemist, complete order is what makes the study of crystals easier, just as, at the opposite extreme of the organization spectrum of matter, complete randomness does the same for gases. The liquid state, where molecules are neither fully ordered nor fully at random, is extremely difficult to model.

The chemical community was introduced to the world of crystals by the advent

Theoretical Aspects and Computer Modeling Edited by A. Gavezzotti

of X-ray crystallography, with its wealth of chemical and structural information. Since, however, the principles and the underpinnings of this discipline are somewhat more complicated than those of many other analytical techniques, the non-crystallographic literature shows sometimes a certain looseness in the use of crystallographic terms, especially in the description of crystal symmetry; it is not unusual, for example, to see the interchange of terms like 'space group', 'structure' and 'lattice', or, even worse, 'cell contents' and 'asymmetric unit'. A few words on the definition of such crucial concepts may be usefully spent here, aiming at chemical understanding rather that at mathematical rigor.

All crystals have translational symmetry, that is, repetition of motifs by rigid displacement in space. The unit cell is the parallelepiped containing the smallest complete structural unit which, when repeated in space in such a way, reproduces the whole crystal structure; several choices are sometimes possible. The vertices of these parallelepipeds constitute a grid of points, of no chemical significance in general, which is called a lattice, with its own periodicity (the edges of the unit cell) and symmetry. Within the cell, symmetry operations relate the structural units which constitute the unit cell content; the asymmetric unit is the smallest structural unit within which no symmetry elements operate. The space group is the collection of the symmetry elements belonging to the crystal structure; it is a group in the mathematical sense, in that each product of two elements must be another symmetry element of the group. The number of symmetry elements in a space group must be equal to the number of asymmetric units in the cell, which, all together, constitute the cell content.

Symmetry operations reproduce chemical entities in space. To define such entities, from now on, the term 'molecule' denotes a molecule proper, or any other chemical entity recognizable also in the gas phase (a helium atom, an SO_4^{2-} ion, an $Fe_2(CO)_9$ complex). In general, it can be said that a molecule is a distinguishable entity when the forces acting within it are much stronger than the forces acting on it in the crystal; however, the definition of the basic chemical unit in a crystal is arbitrary, and, for example, it is usually convenient to consider a molecular complex (say, a charge-transfer bimolecular complex) as a single supermolecule when dealing with the symmetry of its crystal. Difficulties arise with infinite strings or layers; diamond and NaCl crystals are examples of three-dimensionally infinite systems where the term 'molecule' is meaningless. Also, the term 'organic' refers here to crystals containing organic and the large majority of organometallic compounds.

If the molecule has no internal symmetry, then the asymmetric unit is the whole molecule; if two identical molecules join in a dimer without internal symmetry, then the asymmetric unit is two molecules. If a molecule has an internal (point-group) symmetry element which coincides with a space-group symmetry element, then the asymmetric unit is a fraction of a molecule. The following hypothetical examples may be helpful: consider naphthalene, and space group $P2_1/c$, with four symmetry elements. Then:

(a) no molecular symmetry in the crystal: asymmetric unit one molecule, cell content four molecules;
(b) a non-symmetrical dimer of two naphthalene molecules in the crystal: asymmetric unit two molecules, cell content eight molecules;
(c) centrosymmetric molecule in the crystal, center of symmetry coinciding with the crystal center of symmetry: asymmetric unit half a molecule, cell content two molecules;
(d) one molecule without symmetry, one molecule with a center of symmetry coinciding with the crystal center of symmetry: asymmetric unit one and a half molecules, cell content six molecules.

If naphthalene could pack in a space group which contains both a center of symmetry and a mirror plane perpendicular to the molecular plane, and if that molecular symmetry were preserved in the crystal, the asymmetric unit would be a quarter of a molecule, and so on. Atoms lying on the mirror plane would be said to lie on a special position.

It may be surprising to the general chemist that a molecule like naphthalene may or may not have a center of symmetry in the crystal. Experimental evidence tells that organic molecules retain or give up some or all of their point-group symmetry in the crystal, according to the packing pattern, that is, to the energetic requirements of the crystal potential field. In fact, naphthalene takes option (c), retaining its center of symmetry but not its mirror planes and twofold axes. Relationships between point-group and space-group symmetry have been the subject of much speculation in the literature on organic crystal packing. Some of these topics will be taken up again later in this chapter.

1.2 Crystal Symmetry: A Primer

Clearly, the symmetry of the lattice is one thing, and the symmetry within the cell is another. Yet, they interact, and a space group in fact reflects both. A tricky point, not immediately understood, is that some symmetry elements, or their combinations, impose restrictions on the shape of the unit cell, and hence of the lattice. The presence of certain symmetry elements determines the length of some cell periodicities: for example, since applying a screw operator twice results in pure translation, a space group with a screw axis with translation x^0 must have a cell vector (or a cell diagonal) $2x^0$ long. Other symmetry elements restrict some cell angles to unique values, like 90° or 60°. Some confusion arises from reckoning the other way around – a cell angle of 90° does not imply the existence of a certain symmetry element. Thus, a crystal having the proper symmetry elements to be orthorhombic will have all cell angles equal to 90°, but all cell angles being equal to 90° does not qualify a crystal as orthorhombic (even though a non-orthorhombic crystal with such cell angles would be a funny coincidence indeed). The following is a tentative illustration of these concepts.

Most often, the beginner in crystallography learns that there are several crystallographic systems, triclinic, monoclinic, orthorhombic, etc., for which textbooks state the basic laws without much justification: in the triclinic system, cell vectors can have any direction and modulus, while, for example, in the tetragonal system two cell axes must have equal length, and all cell angles are constrained at 90°. Many crystallography teachers and supervisors usually just produce the *International Tables for X-ray Crystallography*, where the beginner is confronted with other dogma. The unique space groups number 230, while the unique lattices are just 17, which is confusing if one has not been told what is the difference between a lattice and a crystal. Only after a little practice is the right conclusion arrived at: molecules arrange themselves in crystals in such a way that the whole spatial arrangement must belong to one of the 230 space groups, while the 17 possible lattices (or periodic arrays of immaterial points) may be filled with symmetry-related molecules in different ways, the number of independent possible combinations being exactly 230 (for a poetic touch, solid-state physicists say that matter 'decorates' the lattice). Only very seldom is the beginner organic crystallographer told that he or she need not worry about 230 space groups, only a handful of them being chemically significant.

Starting crystallography with space groups is starting from the wrong end. Only after a while does the student realize that space groups are made of combinations of symmetry elements; on the other hand, a consideration of symmetry elements in the first place is much more effective. In organic crystallography, the relevant symmetry elements are the inversion center (I), the glide plane (G), the twofold screw axis (S) and, of course, pure translation (T), always present in a crystal. Starting from the basic unit, a molecule, the effect of each of these operators in building space groups will now be described.

The simplest case is when there is no symmetry at all, except translation (Figure 1.1). X and Y belong to some cartesian orthogonal reference frame (say, the inertial axes of the molecule), while the unit cell vectors will be called **a**, **b**, **c**, and the cell angles α, β and γ (these being the angles between the **bc**, **ac** and **ab** directions, respectively). There is complete freedom of translation in all three directions in space; therefore, the **a**, **b**, **c** directions (and the length of the cell edges) are free from constraints. It follows that cell angles also are unrestricted: the system is triclinic, and the space group is $P1$ with one molecule per unit cell. What dictates the length of the unit translations and the angles between the translation directions are just the chemical forces acting between molecules in the crystal, the intermolecular interactions. Adding a constant translation to all atomic coordinates in the molecule simply means that the whole crystal is shifted, undeformed, by the same amount; its total energy is unchanged. The space group is said to be polar in all directions.

An inversion center (I) generates a pair of molecules to be translated in space to form the crystal (Figure 1.2). Not much changes with respect to the previous case, except for one important thing: a center of symmetry now appears at the

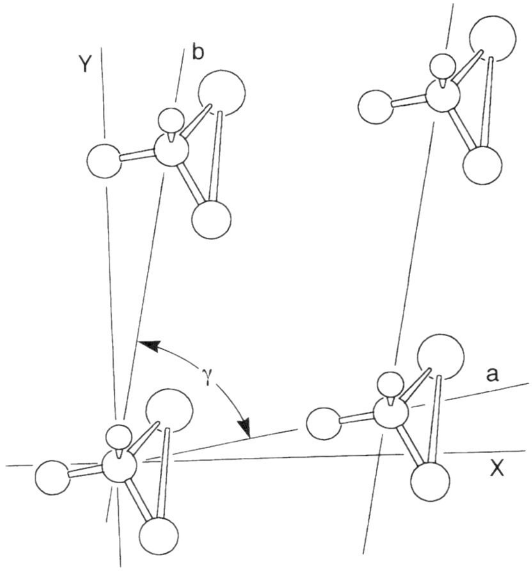

Figure 1.1 A molecule is repeated in space by pure translation. The same sketch applies to *XZ* and *YZ* planes; triclinic system, space group *P*1

midpoint of each unit cell vector, and also in the center of each coordinate plane. Thus, molecule 2 is generated from molecule 1 by *I*, molecule 3 is generated from molecule 1 by *T* and from molecule 2 by *I*. This ensures that the structure is periodic and symmetric. The unit cell vectors can have any direction, so the new centers of symmetry can be anywhere, without restrictions; a center of symmetry inverts all three coordinates anyway. Again, the system is triclinic, and the space group is $P\bar{1}$ with two molecules per unit cell. If the 'dimer' in Figure 1.2 is in fact a single, centrosymmetric molecule, then the space group is $P\bar{1}$ with one molecule per unit cell (half a molecule in the asymmetric unit).

This space group is not polar. Shifting molecule 1, say, along **a**, shifts molecule 2 by the same amount in the opposite direction, under the action of the *I* operator. The two molecules then get closer or further apart, and the total energy does change.

The basic geometrical elements of the arrangement around a twofold screw axis are sketched in Figure 1.3. Starting from molecule 1, molecule 2 is generated by 180° rotation around the screw axis plus translation along *y*. The degrees of freedom in this setup are (i) the orientation of the starting molecule with respect to the *S* axis, (ii) the distance from the center of mass to the *y* axis, which for convenience is taken along *z* (and is called z^0), and (iii) the translation

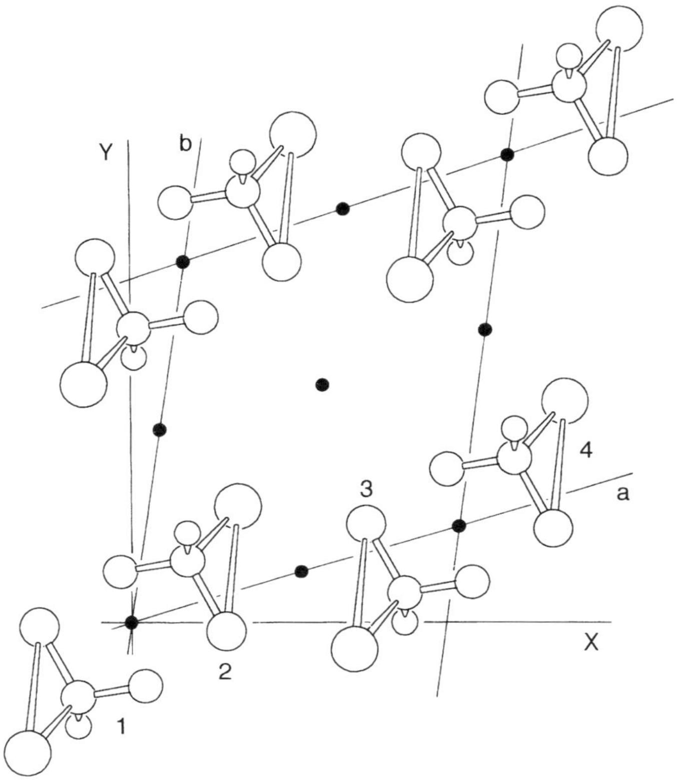

Figure 1.2 Same as Figure 1.1, but for a molecular pair over an inversion center. Triclinic system, space group $P\bar{1}$ (inversion centers are denoted by black dots)

along y (called y^0). Double application of S to molecule 1 (or application of S to molecule 2) gives molecule 3, translated along y by $2y^0$; therefore, the length of the **b** cell axis is fixed, $b = 2y^0$. Again, intermolecular forces alone dictate the optimum geometry, that is, the best orientation, and the best values of y^0 and z^0.

The motif obtained in this way is an infinite string of molecules, which can be translated along x and z; assuming that **b** stays perpendicular to the **ac** plane. Figure 1.4 shows that there is no restriction to the **a** and **c** directions, and hence no restriction on β. But why is it exactly that **b** has to be normal to the **ac** plane, that is, $\alpha = \gamma = 90°$, the monoclinic dogma? Figure 1.5a shows what happens if **a** is taken out of the **ac** plane: molecule 2 is related to 1 by S, but there is no simple symmetry relationship between molecules 2, 3 and 3′. To get to 3 from 2, one cannot use the same S operator: this means that the structure is not symmetric and periodic (the same reasoning holds for the **c** cell axis).

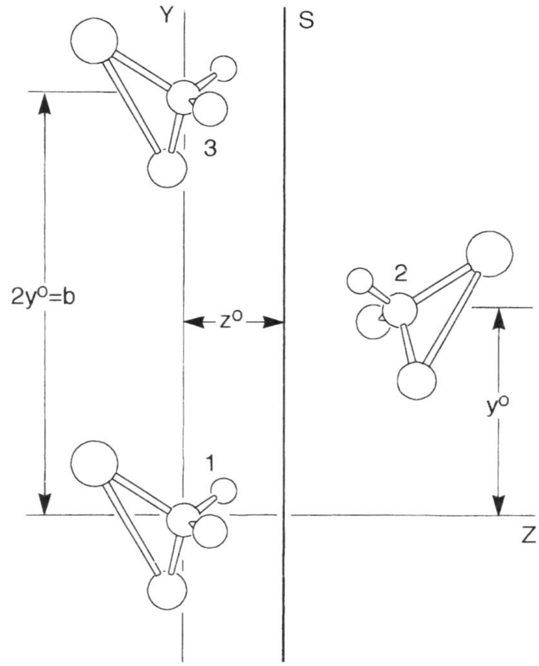

Figure 1.3 The geometry for the action of a screw axis (denoted by *S*). See text for discussion

Figure 1.5b shows the correct geometry: if $\gamma = 90°$, everything falls into place and molecule 3 is connected to 2 by a new screw axis, appearing as a product symmetry at the midpoint of the **a** translation parameter. If the packing forces drive the molecular assembly to adopt an *S* symmetry element, then the resulting crystal must be monoclinic. Since a twofold axis (*A*) is just a screw axis deprived of *y* translation, or formally with $y^0 = 0$, the same reasoning applies. The resulting space groups are $P2_1$ or $P2$ for *S* and *A*, respectively. Both are polar along *y*, but of course in $P2$ the length of the **b** cell vector is also unrestricted.

The setup for the construction of a molecular string for a glide plane perpendicular to *y*, with translation along *z*, is shown in Figure 1.6. Starting from molecule 1, molecule 2 is generated by reflection across *G* plus translation along *z*. The degrees of freedom are the molecular orientation, the distance from the center of mass to the glide plane (y^0), and the length of the *z* translation (z^0). Double application of *G* to molecule 1 (or application of *G* to molecule 2) gives molecule 3, translated along *z* by $2z^0$, so that the **c** cell parameter is fixed, $c = 2z^0$.

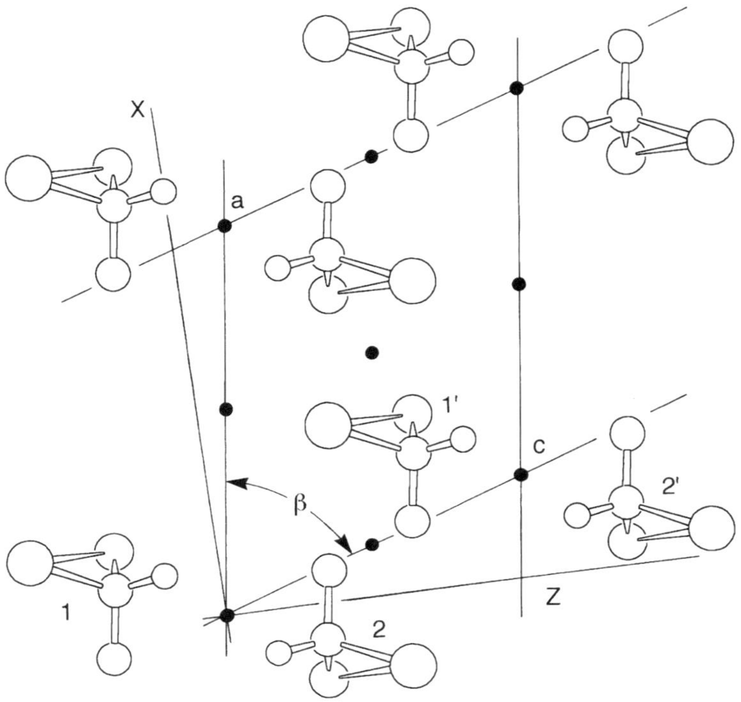

Figure 1.4 The string of molecules in Figure 1.3 is translated in two directions (note the different projection). The screw axes are perpendicular to the *XZ* (**ac**) plane, and their intersections with this plane are denoted by full circles. Space group $P2_1$

Figure 1.7 is the analog of Figure 1.4 for the *G* operator. Here, however, **c** must be along the *z* axis, otherwise the *G* operator cannot survive; remember that a glide (or mirror) operator changes the sign of only one coordinate. The *yz* plane coincides with the **bc** plane, and the **a** direction can still be arbitrary, so β can take any value. The space group is *Pc*, with two molecules per unit cell; it is polar along *x* and *z*, and the system is monoclinic. In this case, and in the case of Figure 1.4 as well, β is arbitrary, including the value of 90°, which in such cases has no special meaning at all. So it is possible to have a monoclinic crystal with $\beta = 90°$, while the crystal symmetry is still monoclinic and need not be upgraded to orthorhombic.

Figure 1.8 is the analog of Figure 1.5 for the *G* case. The structure is not periodic and symmetric when the **b** axis does not coincide with *Y* ($\alpha \neq 90°$, Figure 1.8a), while the correct structure has $\alpha = 90°$ (Figure 1.8b). A new glide plane appears at mid-cell (G').

All the above also explains why Y is called the unique axis in these monoclinic settings. Of course, there being nothing physically unique to any of the cartesian axes, appropriate permutations produce the same space group in different settings and with different unique axes.

When two symmetry elements couple, the offspring is another symmetry element. This is best illustrated in the case of space group $P2_1/c$ (see Figure 1.9). Starting with an I pair (molecules 1–2), and then adding an S operator along Y, the screw generates a string of pairs along Y (molecules 3–4 and 5–6), and the **b** periodicity is, as usual, fixed at $2y^0$ (refer to Figure 1.3). It is easily verified that a G operator appears at $y^0/2$, with a glide translation of $2z^0$ along Z, relating molecules 1–3 and 2–4. The coupling of I and S has generated a new fixed periodicity, twice the glide translation, or $c = 4z^0$. Two unit cell edges having been fixed, a layer has been obtained. Then **a** must be in the xz plane,

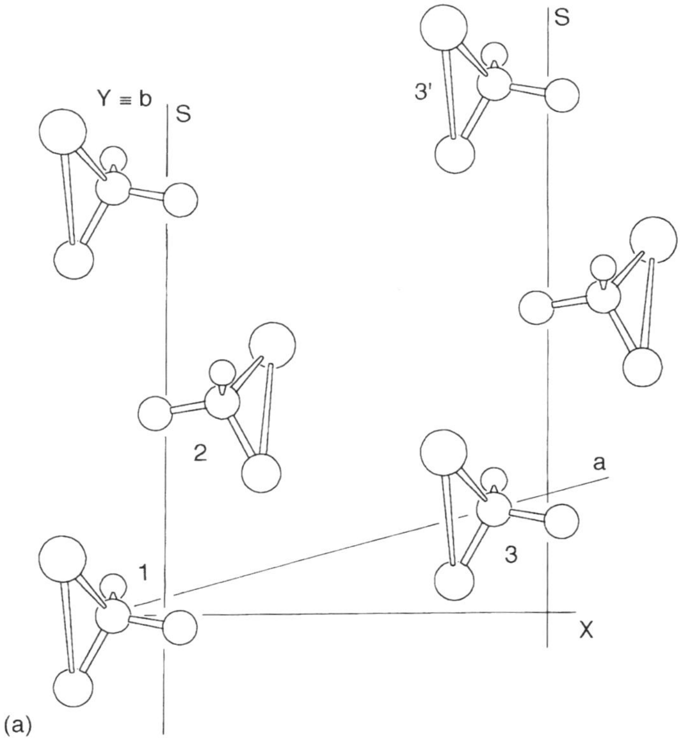

Figure 1.5 (a) The screw axes along Y are not perpendicular to the **ac** plane. The structure is not periodic, molecules 3, 2 and 3′ are not related by screw symmetry. (b) X coincides with **a**, the structure is periodic, molecules 3, 2 and 3′ are related by screw symmetry (S')

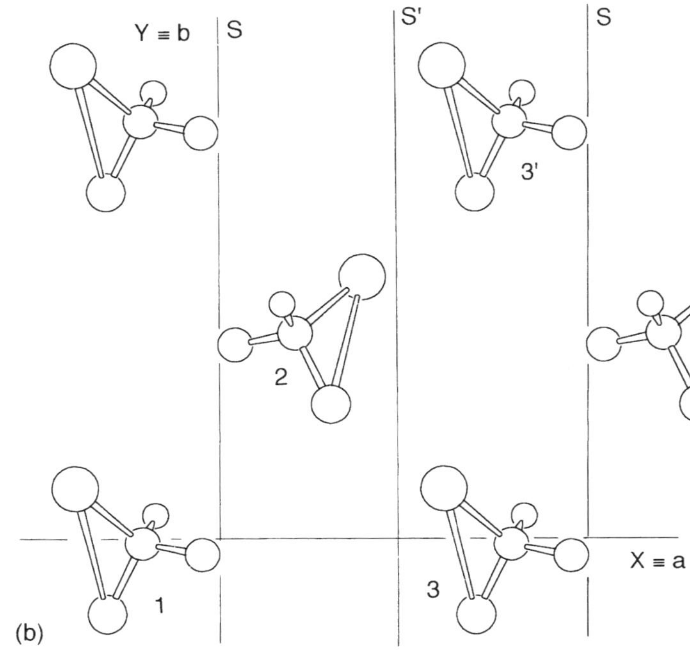

Figure 1.5 (*continued*)

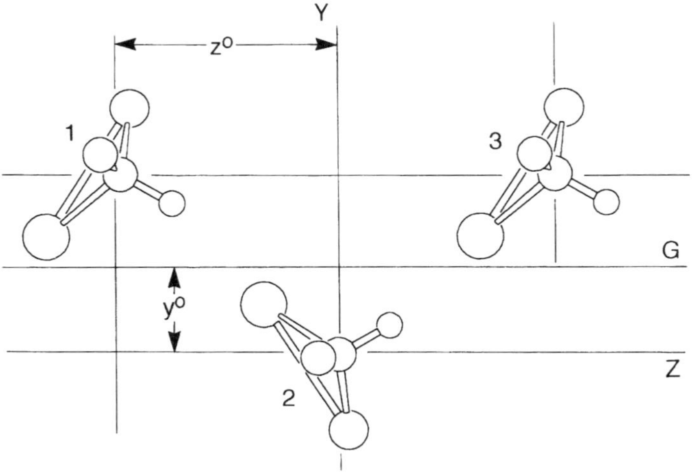

Figure 1.6 The geometry for the action of a glide plane coinciding with the *xz* plane, whose trace is denoted by *G*. See text for discussion

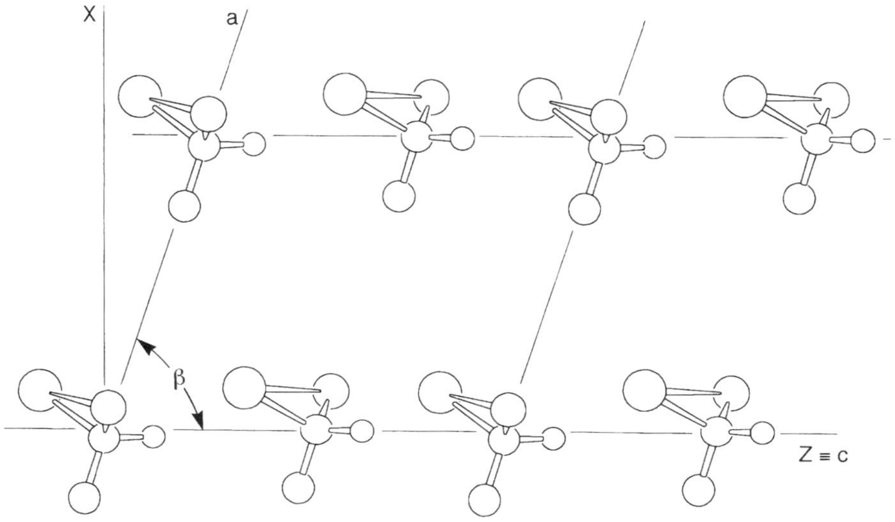

Figure 1.7 The string of molecules in Figure 1.6 (note the different projection) is translated along **a**. The glide plane is parallel to the *XZ* plane. Space group *Pc*

since the glide is parallel to that plane (remember Figure 1.7). No other restrictions apply; the layer repeats itself above and below, but molecules need not be exactly on top of each other, since β need not (although it can) be 90°. This provides further freedom in the crystal packing. The same reasoning could be used for the coupling of any two operators of *I*, *S* and *G* to generate the third one.

Since this space group has four symmetry elements, there are four molecules per unit cell in the former construction, but, if the centrosymmetric 'dimer' is in fact a single centrosymmetric molecule, then the number of molecules per cell is two. It is easy to see why this space group is by far predominant in organic crystallography (about 35% of all organic crystal structures); it comprises all the most favorable symmetry operators (*I*, *S*, *G*) for the packing of organic molecules.

The easiest way of generating orthorhombic symmetry is to start with a screw string along *Y* (like the one in Figure 1.3) and place a second screw axis along *Z*, intersecting the *XY* plane at x^0, y^0. This last symmetry elements repeats in space the whole S_y string, and, in doing so, generates a new screw axis, orthogonal to the first two, and hence along *X*. There is no more freedom: if one screw axis constrains the other two cell directions in the plane perpendicular to it, then understandably three mutually orthogonal screws constrain all cell angles to 90°. The **b** periodicity is fixed by the first screw along *Y* (and not by the value of y^0), the **c** periodicity is fixed by the second screw along *Z*, and the **a** periodicity is

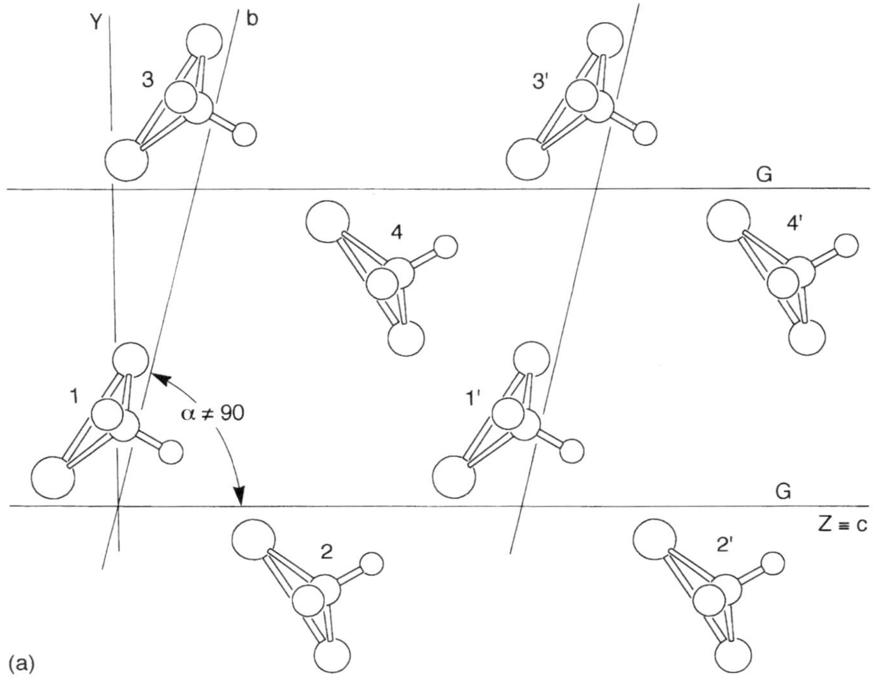

Figure 1.8 As in Figure 1.5, but for the glide plane. (a) **b** does not coincide with *Y*, the structure is not periodic, molecules 1, 4, 1′ are not related by glide symmetry. (b) **b** coincides with *Y*, the structure is periodic, molecules 1, 4, 1′ are related by glide symmetry (G')

fixed at $4x^0$. A three-dimensional symmetric and periodic structure has been generated, and the space group has the most appropriate name of $P2_12_12_1$. Since it contains no centers of symmetry or glide planes, it is a favorite of resolved chiral molecules.

A similar reasoning leads to recognition of other restrictions in systems of higher symmetry. For example, the generation of a periodic ordered structure with a fourfold axis is possible only if two cell edges are constrained to the same length, and all angles to 90° (the tetragonal system). Figure 1.10 shows that after translation along X and Y of a fourfold-symmetric quartet of molecules (1–4) a new fourfold axis appears at the center of the **ab** face, ensuring periodic symmetry to the layer, but only if the two translation vectors have the same length; otherwise the square formed by the centers of mass of the central fourfold-symmetric quartet (molecules $1', 2, 3''', 4''$) would deform into a rhombus. A number of additional twofold rotation axes appear. The length of the Z

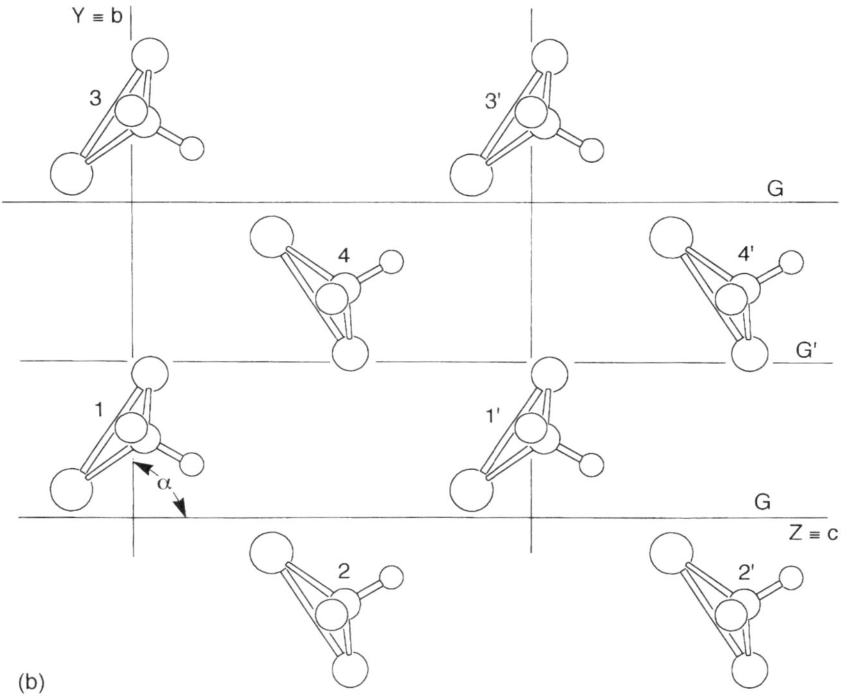

Figure 1.8 (*continued*)

translation is arbitrary, but further consideration of the drawing should provide evidence that all cell angles must be 90°. In a similar way, a threefold or sixfold rotation axis would impose two equal cell edges and one cell angle of 60° (rhombohedral system); and so on.

There is further crystallographic dogma: no fivefold or sevenfold symmetry in crystals. Some textbooks demonstrate this by showing that pentagonal or heptagonal tiles cannot cover a plane without leaving some empty space. Many examples convey a wrong idea, but this example has the unique feature of conveying two wrong ideas, namely, (i) that molecules completely fill the crystal space (this is not the case, the filling coefficient being about 0.7), and (ii) that a molecule with fivefold symmetry will never be able to crystallize (this is not the case, there being a vast literature on the crystal structures of ferrocenes). A more proper way of reasoning is to sketch a picture similar to the one in Figure 1.10, using a fivefold axis, and to try and build a periodic symmetric structure. There

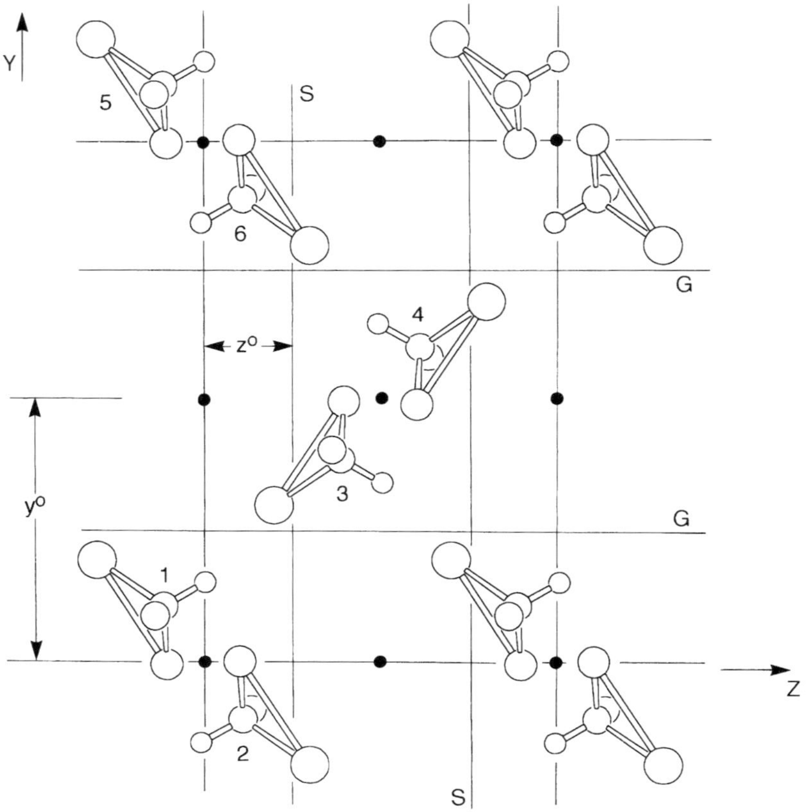

Figure 1.9 The coupling of an inversion center (full circles) with a screw axis (*S*). *G* is the product glide plane. Molecule 1 is related to molecule 2 by *I*, to molecule 3 by *G*, to molecule 4 by *S*. As a whole, the 1–2 dimer is related to the 3–4 dimer both by *S* and by *G*. Space group $P2_1/c$

is simply no way in which molecules related by the first fivefold axis can all be related to all molecules related by a second fivefold axis after translation, so that constant, periodic repetition is impossible. Otherwise, nothing prevents a molecule with fivefold symmetry from packing in space group $P2_1/c$; but the crystal structure has, of course, no fivefold symmetry.

Table 1.1 is a summary of the structural elements (pairs, symmetric clusters, strings, layers, or full three-dimensional structures) that can be obtained by single or multiple application of *I*, *S* and *G* symmetry operators. This table comprises all the most populated space groups for organic molecules.

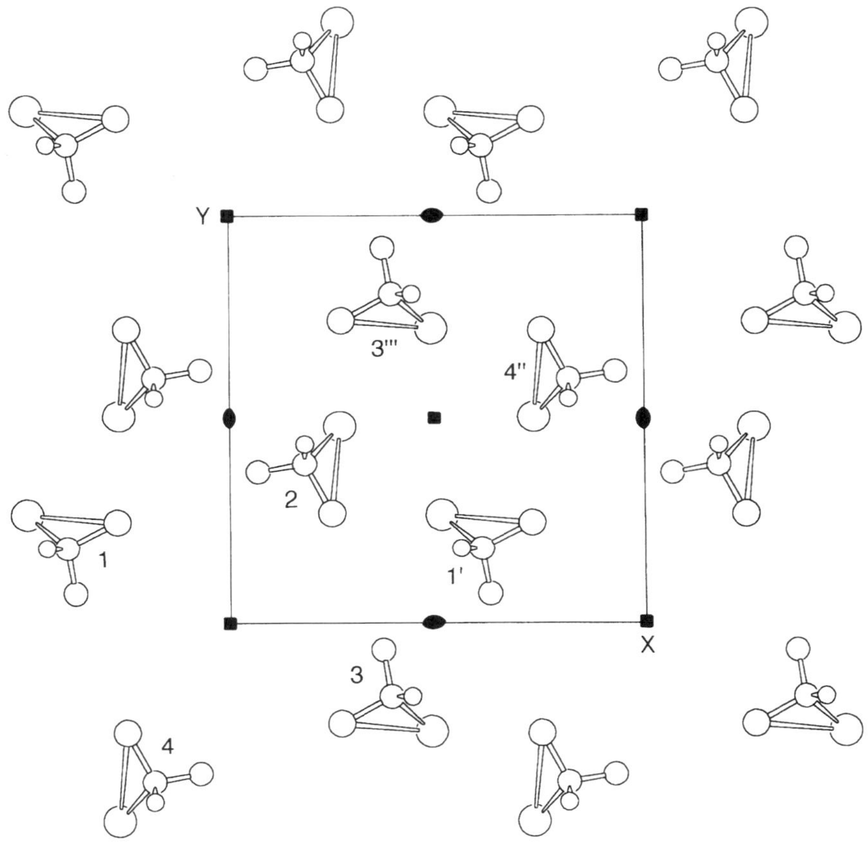

Figure 1.10 A fourfold-symmetric quartet (molecules 1–4) is translated along *X* and *Y*. Fourfold axes are denoted by squares, twofold axes by elliptical dots. Space group *P*4

1.3 Building Model Crystal Structures in a Computer

The line of reasoning taken in the above approach to the construction of space-group symmetry has been exploited in a computer program that reads in a rigid molecule and applies in turn the relevant symmetry operators in organic crystal chemistry, combining them appropriately to reach the most populated space groups. Total energies can then be calculated by some empirical parametric function and compared to each other to find the most promising crystal structures for the given molecule. When applied to favorable cases, this approach has shown a certain predictive ability (see Section 2.5 on suggested reading at the end of this chapter).

Table 1.1 Each square contains the structural units (C = cluster, R = row, L = layer, 3D = full 3D structure) that can be obtained by coupling two symmetry elements (2 = twofold axis, M = mirror, G = glide, S = screw, I = inversion center, CE = centering)

	2	M	G	S	I
2				$P2_12_12$ L	$P2/m$ C $P2/c$ R
M					$P2_1/m$ R $P2/m$ C
G				$P2_1/c$ L $Pca2_1$ L $Pna2_1$ 3D	$P2_1/c$ L $P2/c$ R
S			$P2_1/c$ L $Pca2_1$ L $Pna2_1$ 3D	$P2_12_12_1$ 3D $P2_12_12$ L	$P2_1/m$ R $P2_1/c$ L
I	$P2/m$ C	$P2/m$ C $P2_1/m$ R	$P2_1/c$ L	$P2_1/c$ L	
CE	$C2$ L	Cm L	Cc L		

1.4 Crystal Symmetry: A Matter of Opinion?

Geometrical and mathematical demonstrations on symmetry matters do not take into account experimental uncertainties. In crystallographic practice, cell dimensions and space-group symmetry are usually determined by single-crystal X-ray diffraction, from the spacing and systematic absence (extinction) of certain reflections. Although X-ray diffraction is in this respect an extremely sensitive technique, the choice of a space group is to some extent arbitrary. For example, extinctions say nothing about the presence of a center of symmetry, which is sometimes confirmed only after a careful (but still somewhat arbitrary) refinement of the whole structure. The presence or absence of a symmetry element may depend upon minor details in molecular geometry: the displacement of just one atom from its fully symmetrical location by a few tenths of an angstrom may or may not be enough for X-rays to detect the symmetry violation. Thus, the assignment of a symmetry element may be a matter of sensitivity in the apparatus for the detection of weak reflections (where should be the borderline between a 'very weak' and an 'extinct' reflection can even be a matter of personal taste). From the chemical standpoint, a gross crystal symmetry, insensitive to minor molecular displacements, may be just as appropriate as the much more detailed, or even over-detailed, X-ray crystal symmetry.

Theoretical chemists instinctively consider a crystal as a list of atomic coordinates, where the asymmetric unit is repeated to infinity by space-group symmetry operators. This is the ideal model of a perfect, infinitely ordered crystal, but the real thing is far from that. Even a well-grown single crystal of the size of a few tenths of a millimeter will incorporate dislocations, domains, impurities and other types of structural disorder. In fact, the success of X-ray diffraction itself depends on a balanced distribution of domains, since an ideally perfect crystal would lead to zero net diffracted intensity as a result of multiple interference between diffracted beams. In addition, many organic crystals show several types of structural disorder at the molecular level, leading to several kinds of deviation from perfect Bragg diffraction, like diffuse scattering; this is in itself the subject of a specialized branch of X-ray crystallography.

The size of ideally ordered domains in a real crystal varies according to many factors (chemical constitution, growth conditions, impurities, temperature and pressure). A certain concentration of structural defects is physiological, and may even contribute to lowering the free energy of the crystal; these may be overlooked even in accurate single-crystal X-ray analysis, whose sensitivity threshold towards their detection is of the order of a few percent in concentration, very high if compared with spectroscopic techniques.

Symmetry in real crystals is a sort of macroscopic property, and its platonic statements should be taken with a grain of salt when dealing with molecular-level phenomena. The formulation of a proper theory for the quantitative study of defect distribution in organic crystals is extremely difficult. Apparently, however, thermodynamic properties like the melting point or the melting enthalpy are rather insensitive to the intimate texture of the specimens, as demonstrated by their reproducibility over samples of different origin and morphology. On the other hand, optical or magnetic properties may be largely dependent on defect and impurity concentration. Defects have a much greater influence on electronic than on thermal (vibrational) processes in crystals.

2. MOLECULAR RECOGNITION

2.1 Some Basic Facts

The solid state is, in many respects, easily characterizable. One mole of gaseous substance occupies about 24 liters at room temperature, while the volume of the same amount of substance in the solid state is a few tens to a few hundreds of cubic centimeters. In condensed phases the average volume per molecule is about one and a half times the volume of the molecule itself. Molecules are tightly packed in space, cannot be pushed further against each other, and the compressibility of condensed media is very small. On the other hand, crystals can usually be left in the open air at room temperature for a long time without apparent change in size and shape: besides repelling each other at short distances, molecules in a solid must be held together by some sort of attraction.

Two basic quantitative questions arise: (1) What is the nature and magnitude of the forces between molecules? (2) What is (or are) the favorable geometrical arrangement(s) of molecules at close contact? (The plural here applies to the widespread phenomenon of crystal polymorphism.) These questions can be taken as a definition of the subject of crystal packing. Molecular recognition is, more properly, the detailed mechanism that leads molecules in the gas or liquid (melt or solution) phases to join together in an organized and symmetrical solid. Packing forces and crystal symmetry determine the chemical and physical properties of crystalline materials.

Phase transitions offer important windows looking over this landscape. The melting temperature of a crystal is a sharply defined quantity, often used for characterization. According to thermodynamics, melting occurs with rising temperature, when the entropy gain in the system by spatial randomization overcomes the enthalpy loss due to breaking the crystal-packing forces. When temperature decreases, then, crystallization should occur at the melting temperature, but liquids that behave that way are the exception rather than the rule; in spite of the principles of thermodynamics, crystallization mostly occurs at lower temperatures (supercooling). Similarly, it is usually much easier to dissolve a perfect crystal in an appropriate solvent than to grow back a good crystal from the solution. The obvious conclusion is that a crystal is more readily destroyed than it is formed, or that the molecular mechanisms of nucleation and growth of crystals are under kinetic, rather than thermodynamic, control.

Molecular recognition in fluids far from the melting point averages thermally in a relatively short time, so that equilibrium thermodynamics is adequate for their description. On the other hand, when the pre-crystal stage is approached the use of equilibrium thermodynamics leaves much to be desired. In this stage, small aggregates of molecules are presumably formed, and, since organic molecules have complicated shapes and many possible recognition sites, many competing geometrical structures may be possible. Unpredictable, nearly random local nucleators, like trace impurities or even kinks in the growth vessel, may contribute to widen the spectrum of geometrical and energetic choices. Once formed, these precursor aggregates may build up a considerable activation energy against disruption, so that equilibrium is an optional or at least a distant future (in the molecular time-scale) event. In solution, the solvent plays a key role, as the solute–solute molecular recognition competes with the solute–solvent one. Surface phenomena are crucial.

A large majority of crystal structures apparently do obey the requirements of thermodynamics, eventually reaching the minimum free-energy state. Nevertheless, a true understanding of molecular recognition processes at nucleation stage must require a time-dependent, molecular-level, and hence kinetic approach. Needless to say, kinetics plays a leading role in supercooling or polymorphism. Therefore, a proper theory of crystal precursor states should incorporate a blend of thermodynamics and kinetics: such a theory, as of today, does not exist. The only viable option is apparently the dynamical simulation of small molecular aggregates in solution – a sort of brute force, blindfold approach; still, present-day computational techniques appear somewhat rudimentary against such a formidable task. Such simulations are quite often performed on molecular recognition in the formation of supramolecular complexes, but, not surprisingly, no attempts at performing such simulations with the explicit aim of reconstructing the pre-crystallization stages in a liquid or solution have so far been reported. This seems therefore a most promising direction for future developments.

2.2 The Ingredients of Molecular Recognition

The quantum-chemical foundations of intermolecular interactions are essentially the same as those of ordinary valence; the difference lies in the relative magnitude of the quantum effects involved. A molecule consists of a collection of positively charged atomic nuclei surrounded by an electron cloud. The Hamiltonian for an intermolecular Schrödinger equation contains the same terms as the one for valence interactions; however, intermolecular interatomic distances being larger than valence ones, terms that depend on orbital overlap play a much less overwhelming role. Coulombic terms are responsible for the molecular electrostatic field, which, when polar bonds are present, usually yields the leading terms in intermolecular interactions. However, other crucial effects in molecular recognition, especially for less polar molecular fragments, come from the fact than an approaching charge can polarize, by its own electrostatic field, the electron distribution in a neighboring molecule. Thus, polarization and electron correlation, often neglected in calculations of valence forces, must be taken into account (see Chapter 2).

The attractive forces holding molecules together are a consequence of electrostatic, polarization and electron correlation effects. For organic molecules, the quantitative calculation of these forces has been carried out in a number of different approximations, as described in detail in other chapters in this book. At one extreme, non-polar hydrocarbon molecules must rely on mutual polarization; the resulting forces, usually called dispersion or van der Waals forces, are feeble and non-directional, and may be satisfactorily modeled by entirely empirical potentials. At the other extreme, the hydrogen bond is a strong, directional, almost valence interaction which is the most effective means of recognition between molecules, in organic crystals and in all the biological world as well; for the present purposes, it is proper to state that molecules with donor and acceptor groups will without exception form hydrogen bonds in crystals.

Repulsion at short intermolecular distance can be described in different terms. On purely quantum-mechanical grounds, Pauli's principle states that electrons with the same quantum numbers, no matter that they belong to different molecules, cannot occupy the same region of space. Thus, Pauli 'forces' –

although they are not forces in the sense of Newtonian mechanics – steer electrons to mutual avoidance, forbidding interpenetration of electron clouds. Sidestepping such basic theoretical issues, repulsion is invariably modeled in crystal potentials by steeply rising exponential terms. Repulsion is a basic driving force in the organization of matter. While it can be argued whether or not one attractive arrangement is better than another, there is no question that no repulsive arrangement is permitted, so the optimization of attraction can start only after the requirement of avoiding repulsion has been satisfied.

Unfortunately, this requirement is necessary but not sufficient, there being many different ways in which it can be satisfied, and therefore it has intrinsically little predictive power. Much in the same way, while the physical principles of intermolecular attraction are not particularly unusual, the way they apply to crystal nucleation, crystal growth and phase behavior is of great complexity both in space and in time. This is the reason why present-day theories on these phenomena are still in a rudimentary state.

2.3 Symmetry, Order and Crystal Stability

Intermolecular attraction brings molecules together, but there is no *a priori* implication of order and symmetry. Figure 1.5 illustrates on purely theoretical grounds the necessary environment for obtaining a periodically symmetric structure, but is far from demonstrating the energetic necessity of such a structure. Glasses, where molecules are oriented at random, are sometimes as stable as crystals. The following sections are a discussion on some aspects of the relationships between symmetry and stability, but the reader should be warned that necessity in such a context has often to be replaced by much weaker concepts, like inference or analogy.

A few preliminary considerations concern chirality. Chirality is a sort of pre-symmetric property, both in crystals and in free molecules. Space groups that do not contain inversion centers, mirror planes or glide planes are chiral. Solutions or melts of chiral molecules, containing only one optical isomer, must crystallize in a chiral space group; this is the case for many natural products whose crystal structures appear in the crystallographic literature, since they were isolated in optically pure form from plants or animals. Interestingly, however, sometimes a racemic solution or melt produces chiral crystals, in which only one optical isomer appears; this spontaneous resolution by crystallization is often quoted as one of the possible sources of optical resolution in the biological world. It is difficult, if not impossible, to ascertain its frequency in laboratory practice, because the chemical history of the starting materials is almost never mentioned by authors of crystallographic papers. Anyway, the reasons for such a selectivity, related either to the thermodynamics or the kinetics of crystal formation and growth, are far from understood. Comparative studies have not revealed significant or systematic differences in crystal compactness between chiral and achiral crystals [1].

The standard way of discussing chemical stability is in terms of free energy of phases, and of activation energies for phase transitions. Therefore, a chemically orthodox way of dealing with relationships between order and stability is to analyze the connections between crystal symmetry and thermodynamic functions on the one hand, and between crystal symmetry and kinetic constants on the other hand. If the first is a difficult task, the second is an almost impossible one.

2.3.1 Crystal Symmetry and Entropy

There is no contribution to configurational entropy from the symmetry operations in a crystal structure. As long as the crystal structure is ordered, a more symmetric space group has just the same configurational entropy (that is, zero) as a less symmetrical one, as is evident from a consideration of the so-called third principle of thermodynamics; at 0 K, entropy must be zero whatever the space group, and there is no way in which this situation can change with temperature.

The one contribution to the system entropy that is distinctive of a crystal structure comes from lattice vibrations. A subtler connection may exist between the symmetry of a given space group and the vibrational freedom of molecules: a study of the relationship between the chemical constitution of a molecule, the symmetry of its lattice vibrations, and their energy spacings would be a worthy enterprise, but, to this author's knowledge, this relationship has never been quantified or even clearly identified. Matters are made even more complicated by the fact that lattice-vibrational frequencies are variable over the Brillouin zone.

2.3.2 Crystal Symmetry and Enthalpy

Thus, it is more customary to look for connections between crystal symmetry and crystal stability in terms of enthalpy, by an intuitive analysis of the geometry of molecular arrangements against the strength of intermolecular interactions. The task is now restricted to finding those space symmetries that are likely to provide favorable intermolecular exposure.

The contribution to enthalpy that is intrinsic to a crystal structure is the one arising from intermolecular contacts, and symmetry will have to be such as to foster the maximum number of favorable, and the minimum number of unfavorable, juxtapositions. This requirement can be investigated, at least in part, by geometry alone; since the leading terms in intermolecular attraction are the electrostatic ones, the chemical constitution and the geometrical structure of the molecule may offer clues on the best orientations. However, the molecular electrostatic field is complex and diffuse, and localizing the relevant interactions is a dangerous exercise. Judging polarization from geometry is even more problematic. In conclusion, there is no safe way, in general, of tracing back molecular recognition to molecular structure, except perhaps when hydrogen bonding is present. Even in that case, rules can be found for the preliminary act of molecular recognition, the formation of a hydrogen-bonded dimer or ring, but then one is again confronted with the difficulties of the mutual recognition of these new units. In a typical example, it is easily predicted that carboxylic acids will form ring dimers in the solid state, but then the real problem is the packing mode of the dimers.

A less stringent, but safer statement is that empty space is an energetic waste, so that molecules will interlock to reach good space-filling arrangements. This close-packing idea appeared very early in a primitive form, but was consciously put forward by Kitaigorodski [2]. It is sometimes rephrased in terms of bumps-in-hollow fitting, but such a wording may be misleading, as it points to steric avoidance (which is only a preliminary requirement), while the attractive effect is intrinsically electrostatic or even purely quantum-mechanical.

Order and symmetry now come to the foreground, since, for an array of identical objects, periodic, ordered and symmetrical structures naturally lead to a more efficient use of space, and therefore to promotion of favorable contacts. This statement cannot be demonstrated rigorously, but is substantiated by the fact that organic solids are mostly crystals. As mentioned at the beginning of this chapter, this was astonishing to early X-ray crystallographers, and, for that matter, is astonishing even today. Chemists now fully appreciate the fact, but still barely understand it.

Other experimental proof of the space-filling, close-packing principle is the fact that the density of ordered crystals is invariably higher than that of glasses and melts. This observation is probably at the origin of the identification of energetically favorable packing with close-packing; note that the water crystal does not close-pack, as it is an array not of molecules, but of hydrogen bonds. In fact, when particularly strong and directional interactions (like hydrogen bonds) are present, the close-packing requirement might be a little less stringent, since, in principle, these special interactions might be satisfied at its expense. However, it turns out that all crystals, hydrogen-bonded or not, have a packing coefficient between 0.65 and 0.80. The density decrease on melting does not exceed 15%, so that liquids must have packing coefficients around 0.57–0.70. A space-occupancy factor larger than 0.5 is a *sine qua non* for organic condensed phases at room conditions, while the range between zero (a gas) and 0.5 is a no man's land for ordinary condensed matter, and may pertain only to supercritical fluids.

2.3.3 Crystal Symmetry and Space Filling

Unfortunately, the above argument does not work the other way around. All observed crystal structures are close-packed, but not all close-packed structures have been observed, as hundreds of unobserved close-packed structures can be generated by computer for any given molecule. Close-packing is necessary but is neither sufficient nor even indicative, which, admittedly, leaves the close-packing principle as a remarkable insight for a time when no large computers and databases were available, but also as a concept totally deprived of positive predictive power.

It does, however, have negative predictive power, in that it can be used to screen out all crystal structures having a packing coefficient less than 0.65. Experimental evidence over thousands of crystal structures, as well as theoretical analysis, tells us that close-packing can be accomplished only by a limited number of combinations of very few symmetry operations; for organic compounds, those performed by the inversion center (*I*), the twofold screw axis (*S*) and the glide plane (*G*). For objects of irregular shape, mirror planes and twofold axes produce detrimental voids in the supramolecular assembly. As a consequence, some space groups are mathematically legitimate, but chemically insignificant, and the so far determined crystal structures of organic compounds belong to a rather restricted number of space groups.

The role of pure translation (*T*) has probably been underestimated in all studies on the theory of crystal packing. This symmetry operation, the only one that is by definition always present in any crystal structure, is presumably more efficient towards close-packing than is usually thought; space group $P1$ (where translation is the only symmetry element) is the ninth most populated one for organic substances. Had translation been detrimental to close-packing, no crystals as we know them could have been energetically legitimate.

The fact that crystal structures have been observed in all of the 230 space groups has been, half-seriously, quoted as the 'experimental proof of space-group theory' [3]. One could argue whether an axiomatic theory needs any experimental proof; it is true by definition, once it has been made free from trivial mistakes. On the other hand, experimental proof of the assertion that only some space groups are chemically relevant comes from a statistical analysis of the space-group frequencies in organic compounds, whereby the leading role of *I*, *S* and *G* in close-packing comes to the foreground. Care must be taken to distinguish between intermolecular and intramolecular symmetry elements (the latter being the ones that are also point-group symmetry elements; clearly, a mirror or a twofold axis which relate different parts of the same molecule are irrelevant to crystal packing). When point-group symmetry is filtered out, most, if not all of the observed organic crystal structures above orthorhombic are downgraded to the usual pool of 'organic' space groups. Viewing the issue from the other end, Table 1.1 shows that a pairwise combination of *I*, *S* and *G* produces rows, layers or full structures in all the most populated space groups for organics.

Similar statistics, with a recognition of point-group elements, are not available for inorganic compounds, but a similar trend would probably be found. These compounds frequently contain highly symmetrical (tetrahedral, triangular, square-planar) ions or groups, which carry over their symmetry to the crystal. This causes a spread of the space-group frequencies towards the tetragonal, hexagonal or cubic systems (a no man's land for organics). No space group has more than 10% of inorganic crystal structures [3].

One can never be too careful when generalizing on such topics; crystal packing is a subtle, elusive subject. A symmetry operation is energetically relevant only when it relates nearest-neighbor molecules. Wilson [4] has pointed out that, in some space groups, some symmetry elements may be silent, or 'encumbered'; they appear as products of other elements, but are often prevented, by their location in space, from acting between nearest-neighbor molecules. For example, in *Pbca*, a centrosymmetric space group, the inversion center is very seldom relevant to crystal packing [5].

2.3.4 Crystal-Symmetry Weakening

In spite of all this, the self-organization of matter in the formation of solids seems to have a number of ways that defy classification and, above all, dispense with symmetry. A good percentage (~20%) of the observed organic crystal structures have more than one molecule in the asymmetric unit; thus, the basic act in the formation of the crystal, the recognition of molecular partners in the plurimolecular asymmetric unit, takes place in open violation of symmetry requirements. The supramolecular entity thus formed does, however, pack in a symmetrical arrangement. Thus, violation of symmetry can take place at short range, in the local environment which forms a specially stable two- or three-molecule aggregate, but in the long range the whole solid assembly must be periodically symmetric.

Figure 1.5a shows a mathematically unorthodox structure, which is, however, reasonably close-packed and free from repulsive contacts. Such a picture describes the origin of twinning, where two periodically ordered structures coexist, being related by a pseudo-symmetry operation. The energetic necessity of such symmetry-weakening constructions is not understood, but some of them may be entropically favorable (in a configurational sense), and as such may become more and more relevant with increasing temperature. A rather common occurrence is structure-disordering by molecular reorientations, which increase the time-averaged symmetry of the crystal and produce a marked increase of configurational (and presumably also of vibrational) entropy. In the so-called plastic phases, all molecules in the solid reorient in phase, losing their geometrical identity as the structure tends more and more to a close-packing of spheres. The final breakdown of translational symmetry leads eventually to the liquid state. Even there, some preferential orientation, reminiscent of solid-state symmetry, can be retained as a memory effect, if the temperature is not too far from the melting point.

2.3.5 Crystal Symmetry and Kinetics

No representations, analyses or even speculations on the connections between crystal symmetry and the kinetics of phase transformations in solids have ever been presented; either there are none, or they are too difficult to investigate. In any case, the kinetics of transformations in which at least one partner is crystalline has not so far been studied in a systematic way at the molecular level.

2.4 Crystal Symmetry and Molecular Recognition: A Key to Crystal Design?

Molecular recognition has recently risen to the status of frontier topic in crystallography and in the biological sciences. A crystal is a macro-supramolecular assembly where one should more properly speak of molecular self-recognition. The logical consequence of this way of thinking is that a crystal structure should be understood, predicted or even controlled by an analysis of the basic linking points in the molecular structure of the constituent; this exercise, whose intellectual beauty has so far not always been matched by predictive success, is at the core of a newborn discipline called 'crystal engineering'. For the moment, it consists of detailed analyses of existing geometrical data or recognition patterns, the clearest among which are based on hydrogen bonds [6], or of clever analogies with the logic of synthetic organic chemistry [7].

Crystal symmetry is a mathematical and geometrical theory that can be learned with relatively little effort, and the physics of the interaction between molecules can be understood on a standard quantum-chemical basis. The task of putting together these two pieces in a comprehensive theory of crystal packing has not yet been accomplished, and the rules that determine the ways in which intermolecular forces can be satisfied are complex and still obscure.

Denoting by A the basic structural unit (a molecule), in the early stages of nucleation, molecular recognition can conceivably act either by cooperative addition, of the type $n\mathrm{A} = \mathrm{A}_n$, with n of the order of tens or hundreds, or by stepwise addition, of the type $\mathrm{A} + \mathrm{A} = \mathrm{A}_2, \ldots, \mathrm{A}_n + \mathrm{A} = \mathrm{A}_{n+1}, \ldots,$ with n going from 2 to infinity. In the first case, growth must ultimately depend on subsequent stepwise addition. Clearly, the concept of molecular recognition is itself ill-defined in the first case, and is more localized in the second case. Later, each growth nucleus can live an enthalpic, entropic and kinetic life of its own, and at some stage evolve to a more stable structure. Symmetry, which has been seen to be a fundamental aspect of fully grown solids, can be only marginally important in these early stages.

Small molecular clusters are being studied by diffraction experiments and dynamical simulations [6]; however, there is no comprehensive theory of the evolution of these elusive entities, the crystallization nuclei, towards a full crystal, nor have adequate experimental tools been developed for this kind of investigation. A host of experimental conditions, including of course the degree of supersaturation in solution, influence the very complex lifespan of the nuclei.

At the other end of the range, the study of crystals by diffraction and other methods yields a large amount of raw information, whose organization is still very schematic. Crystal symmetry is detected, but its ultimate origin is unexplained, except for a few general principles, some of which have been discussed in this chapter. Geometrical recognition patterns are also detected, catalogued, and painstakingly organized, but still with a great amount of subjective interpretation. None of these principles yet has a truly positive predictive power, and none has yet been cast in the proper language of physical chemistry, thermodynamic and kinetics. Even computer simulations are making their way through thermodynamically and kinetically uncharted land: they are sometimes successful in solving crystal structure in synergy with partial diffraction data, but are generally unsuccessful in real prediction.

For these reasons, crystal-packing prediction and control are still faraway goals. A crystal structure is a diffuse and delicate construction resulting from a complex force field. Singling out partial terms in this field, ranking them in order of importance, and deciding upon their relative balance, are difficult if not impossible tasks, at least in a general sense, and, even if these tasks could be carried out with confidence, no information on kinetics would be made available. In a more concise statement, there are simply too many spatial possibilities with very nearly the same free energy, at least to our eyes today, and kinetic factors are important and even less understood than thermodynamic ones.

In spite of all the above skepticism, it may well be that an empirical mix of geometrical correlation and computer simulation will produce in the near future a practical tool for the *ab initio* solution of organic crystal structures. An expert system handed by a computer, however, lacks the intellectual appeal of a physicochemical theory, and, on a rather narcissistic note, it would be a pity if one of the last remaining mysteries of organic physical chemistry – the relationships between a crystal and its constituents – were to be left unsolved because a run-of-the-mill computer procedure can find ways around it. On the other hand, a proper theory of nucleation, growth and stability of organic crystals should interactively join together bits and pieces of different origin, from fluid dynamics to thermodynamics and kinetics, dealing at the same time with the complexity of starting from a molecule to build a macroscopic object. It is perhaps no wonder that such a theory is not yet in sight.

2.5 Related Literature and Suggestions for Further Reading

The subject of this chapter is so broad that no detailed references were given in the text, except where unavoidable. The following is a review of literature generally related to the topic.

General aspects of symmetry in chemistry and crystallography: some chapters in *Symmetry, Unifying Human Understanding*, edited by I. Hargittai, International Series in Modern Applied Mathematics and Computer Science, Vol. 10, Pergamon Press, New York (1986); E. Heilbronner and J. D. Dunitz, *Reflections on Symmetry*, VCH, Weinheim (1993).

For a rigorous treatment of point- and space-group symmetry, see G. Burns and A. M. Glazer, *Space Groups for Solid State Scientists*, 2nd edition, Academic Press, Boston (1990).

A classic study of the chemical consequences of crystal symmetry is in D. Y. Curtin and I. C. Paul, *Chem. Revs.*, **81**, 525 (1981).

The computer program to build crystal structures from symmetry operators is called PROMET (A. Gavezzotti, University of Milan, 1995). For applications, see A. Gavezzotti, *Acta Cryst.*, **B52**, 201 (1996).

'Fuzzy' symmetry: see H. Zabrodsky and D. Avnir, *J. Am. Chem. Soc.*, **114**, 462 (1995), and references therein. For uncertainties in the assignment of space groups, see R. E. Marsh and I. Bernal, *Acta Cryst.*, **B51**, 300 (1995). Diffuse scattering from disordered crystals is treated in F. Frey, *Acta Cryst.*, **B51**, 592 (1995).

Molecular-dynamics simulations for solutions and supramolecular chemistry: see for instance W. L. Jorgensen, *Chemtracts – Organic Chemistry* **4**, 91 (1991).

Polymorphism: A. Gavezzotti and G. Filippini, *J. Am. Chem. Soc.*, **117**, 12299 (1996).

A reference reading on crystal growth is J. Hulliger, *Angew. Chem. Int. Ed. Engl.*, **33**, 143 (1994). The energetic and computational aspects of crystal growth are discussed in Chapter 7 of the present book.

A compendium of the theory of the structure and of the optical, electrical and chemical properties of organic materials is in J. D. Wright, *Molecular Crystals*, Cambridge University Press, Cambridge (1987). A quick reference monograph on the nature of intermolecular forces is M. Rigby, E. B. Smith, W. A. Wakeham and G. C. Maitland, *The Forces between Molecules*, Clarendon Press, Oxford (1986). The empirical fitting of potential functions for organic crystals and the general topic of crystal potentials is dealt with at length in several chapters of the present book.

On spontaneous resolution by crystallization see, for example, S. Larsen and K. Marthi, *Acta Cryst.*, **B51**, 338 (1995).

Space-group statistics: priority on this topic goes apparently to Nowacki: W. Nowacki, *Helv. Chim. Acta*, **25**, 863 (1942); *Helv. Chim. Acta*, **26**, 459 (1943); see H.-B. Buergi and J. D. Dunitz, *Helv. Chim. Acta*, **76**, 1115 (1993). Mention of uneven distribution among space groups is of course throughout Kitaigorodski's work. For systematic, quantitative studies over crystallographic databases see Ref. [3] in this chapter; C. P. Brock and J. D. Dunitz, *Chem. Mat.*, **6**, 1118 (1994); A. D. Mighell, V. L. Himes and J. L. Rodgers, *Acta Cryst.*, **A39**, 737 (1983); N. Padmaja, S. Ramakumar and M. A. Viswamitra, *Acta Cryst.*,

A46, 725 (1990); R. P. Scaringe, in *Electron Crystallography of Organic Molecules*, edited by J. R. Fryer and D. L. Dorset, p. 85, Kluwer Academic Publishers, Dordrecht (1991). See also V. K. Belsky, O. N. Zorkaya and P. M. Zorkii, *Acta Cryst.*, **A51**, 473 (1995).

The space-group distribution in databases is, however, biased: chiral space groups may appear overpopulated since many natural products are isolated in optically pure form.

Close-packing: W. Barlow and W. J. Pope, *J. Chem. Soc.*, 1675 (1907); Ref. [2] in this chapter; A. I. Kitaigorodski, *Organic Chemical Crystallography*, Consultants Bureau, New York (1961). The concept has permeated the crystallographic literature for years, while its theoretical foundations [2] have been somewhat lost or vulgarized to its more easily grasped consequences (see the critique of the bump-in-hollow concept in Section 2.3.3 of this chapter). See also P. M. Zorkii and M. A. Porai-Koshits, *Soviet Physics – Crystallography*, **12**, 863 (1968); a more recent paper deals with protoaggregates and crystal symmetry: P. M. Zorkii, *Russ. J. Phys. Chem.*, **68**, 966 (1994).

The packing coefficient [2] is the ratio of volume occupied by molecules in the cell to the volume of the cell. Molecular volumes can be calculated in a number of ways; the simplest are in A. Bondi, *J. Phys. Chem.*, **68**, 441 (1964), and in A. Gavezzotti, *J. Am. Chem. Soc.*, **105**, 5220 (1983).

Two fundamental books on the chemical, physical and thermodynamic properties of organic condensed phases are those by A. Bondi, *Physical Properties of Molecular Crystals, Liquids and Glasses*, Wiley, New York (1968), and A. R. Ubbelohde, *The Molten State of Matter*, Wiley, Chichester (1978). They are probably long out of print, but must be still available in your chemistry library.

A recent book on hydrogen bonding in crystals is G. A. Jeffrey and W. Saenger, *Hydrogen Bonding in Biological Structures*, Springer-Verlag, Berlin (1991). See also, for H-bonding recognition in connection with crystal symmetry, A. Gavezzotti and G. Filippini, *J. Phys. Chem.*, **98**, 4831 (1994); C. P. Brock and L. L. Duncan, *Chem. Mater.*, **6**, 1307 (1994).

Molecular recognition, in its classical sense of inclusion or host–guest compound formation, is not the subject of this chapter, which is aimed rather at the specific process of crystal formation. Examples of crystal structure design by control of prearranged hydrogen-bonding units is a flourishing topic in chemistry; see, for example, G. M. Whitesides, E. E. Simanek, J. P. Mathias, C. T. Seto, D. N. Chin, M. Mammen and D. M. Gordan, *Acc. Chem. Res.*, **28**, 37 (1995); A. D. Burrows, C.-W. Chan, M. M. Chowdhry, J. E. McGrady and D. M. P. Mingos, *Chem. Soc. Revs.*, **24**, 329 (1995). On the general aspects of possible control and prediction of crystal structure see A. Gavezzotti, *Acc. Chem. Res.*, **27**, 309 (1994). The computational aspects of crystal structure prediction are discussed in Chapters 4 and 6 of the present book.

Molecular recognition at interfaces has been studied over the years by several groups at the Weizmann Institute of Science. See, for a review, I. Weissbuch,

R. Popovitz-Biro, M. Lahav and L. Leiserowitz, *Acta Cryst.*, **B51**, 115 (1995); J. Aizenberg, J. Hanson, T. F. Koetzle, L. Leiserowitz, S. Weiner and L. Addadi, *Chem. Eur. J.*, **1**, 414 (1995).

A thought-provoking introduction to some vibrational properties of liquids with respect to solids is R. M. Stratt, *Acc. Chem. Res.*, **28**, 201 (1995).

3. REFERENCES

1. C. P. Brock, W. B. Schweizer and J. D. Dunitz, *J. Am. Chem. Soc.*, **113**, 9811 (1991).
2. A. I. Kitaigorodski, *Organic Chemical Crystallography*, Consultants Bureau, New York (1961).
3. W. H. Baur and D. Kassner, *Acta Cryst.*, **B48**, 356 (1992).
4. A. J. C. Wilson, *Acta Cryst.*, **A44**, 715 (1988); *Acta Cryst.*, **A46**, 742 (1990).
5. G. Filippini and A. Gavezzotti, *Acta Cryst.*, **B48**, 230 (1992).
6. G. R. Desiraju, *Crystal Engineering*, Elsevier, Amsterdam (1989).
7. G. R. Desiraju, *Angew. Chem. Int. Ed. Engl.*, **34**, 2311 (1995).

Chapter 2

Intermolecular Forces–from the Molecular Charge Distribution to the Molecular Packing

S. L. PRICE
Department of Chemistry, University College London

1. INTRODUCTION

Molecular crystal structures are generally densely packed, so that Kitaigorodsky's statement [1] that the molecules pack so that 'the bumps in one molecule fit the hollows in another', providing a high degree of steric interlocking, is an excellent starting point for understanding their structures. However, for refining structures suggested by such a mechanical model, or considering alternative relatively close-packed structures, estimates of the lattice energy are required.

The isotropic atom–atom potential has become the basic model used for molecular crystal structure simulations, with its successes being well documented by Pertsin and Kitaigorodsky [1]. It is an empirical approximation that provides a slightly softened spherical repulsive wall around each atom, a general attractive dispersion force to guarantee aggregation, and usually an electrostatic model to represent the intermolecular repulsions and attractions between polar atoms. A more elaborate model was practically impossible prior to the recent advances in computer power. The hope that this simple model would prove an adequate advance on the close packing of hard molecular shapes has been

Theoretical Aspects and Computer Modeling Edited by A. Gavezzotti

justified in some cases. However, it has proved inadequate in others, leading Desiraju [2] to the conclusion that the isotropic atom–atom model 'gives an approximate description but does not stress adequately many structure-defining interactions'. The model could be improved empirically by refining the model intermolecular potential form for specific interactions, such as hydrogen bonds, or Cl···Cl close contacts. However, this empirical approach has the problem that some crystal structures are sufficiently determined by the close-packing principle to be relatively insensitive to the model potential, and so potentials developed primarily on such packings fail for more sensitive structures where a wide range of comparably close-packed low-energy structures are possible. To give an extreme example of this, an intermolecular potential for two X atoms that was satisfactory for a molecule with such a complex shape that only one close packed structure is possible, is unlikely to be successful for the structure of X_2, because of the large range of close-packed structures that are adopted by the homonuclear diatomics. (This is assuming that any potential would be transferable between the two molecules, a point that is discussed in Section 3.3.) This is just one of the reasons why the empirical development of potentials for modelling molecular crystal structures may have reached its limit with the empirical 6-exp potentials of Gavezzotti and Filippini [3].

An alternative approach is to use the best possible model for the intermolecular potential, derived as far as possible from external, more rigorous sources. The entire simulation method, including the practically necessary approximations to the intermolecular potential, can then be assessed by comparing the predicted properties of the molecular crystal with experiment. This approach is only conceivable because of the significant advances in the theory and quantification of intermolecular forces that have occurred in the last decade or so. The next section outlines the current state of the theory of the intermolecular forces between closed-shell polyatomics, giving a few references for access into this large and active field. Application of these theories is gradually being extended from small molecules in van der Waals clusters, to typical small organic molecules. This suggests a range of improvements that would be needed for an atom–atom-based potential scheme to be reliable across a wide range of types of molecules, and the specific terms that would be more important for certain functional groups. The progress to date on reliable potentials for small polyatomics is briefly discussed, mainly to illustrate the many factors that need to be considered for accurate intermolecular potentials for organic molecules (Section 3).

The experience in modelling van der Waals complexes suggests that an accurate description of the electrostatic forces is necessary for a reliable description of the orientation dependence of the intermolecular forces. As such, this is the first step towards placing the model potentials used in crystal-structure prediction on a firmer theoretical footing. The electrostatic interaction can readily be modelled quite accurately, by sets of atomic multipole moments derived from the wavefunction of the molecule, as described in Section 2.3.1. The implementation of such distributed multipole electrostatic models in molecular crystal-structure modelling is outlined in Section 4. This has been used to demonstrate the sensitivity of a wide range of crystal structures to the accuracy and detail of the electrostatic model. The addition of an accurate electrostatic model, which accurately reflects the molecular charge distribution, to an empirical atom–atom repulsion–dispersion model, provides an overall potential which has a minimum in the lattice energy at a structure reasonably close to that found experimentally for a wide range of crystal structures of polar organic molecules.

Although further improvements to the model potential will be made in the future, and will be necessary for some molecules and some properties, these too are likely to be derived from the wavefunctions of the isolated molecules. Hence, application of the theory of intermolecular forces will play a major role in making computer simulation a reliable and predictive tool for studying molecular crystal structures.

2. THE THEORY OF INTERMOLECULAR FORCES

2.1 The Pairwise Additive Approximation

Formally, the total intermolecular energy of an ensemble of N rigid molecules (or lattice energy for a crystal structure as N becomes very large) can be written as

$$U = \sum_{i<j}^{N} U_{ij}(R_{ij}\Omega_{ij}) + \sum_{i<j<k}^{N} U_{ijk}(R_{ij}, R_{ik}, R_{jk}, \Omega_{ijk}) + \sum_{i<j<k<l}^{N} U_{ijkl}(R_{ij}, R_{ik}, R_{il}, R_{jk}, R_{jl}, R_{kl}, \Omega_{ijkl}) + \ldots$$

where the first term is the sum of all two-body (pair) interactions (each a function of the separation R_{ij} and relative orientation Ω_{ij} of the two molecules). The three-body term U_{ijk} provides the difference between the actual energy of a trio of molecules in a given orientation and the sum of the three pair potential terms; similarly, U_{ijkl} is the corresponding correction to give the correct energy of a quartet of molecules; and so on. Most crystal modelling makes the pairwise additive approximation by only using the first term in this expansion. Hence, it must be remembered that even if the true intermolecular pair potential for a molecule were available, the lattice energy evaluated using it would be in error by the non-additive terms. Conversely, empirical pair potentials fitted to condensed phase properties will partially absorb some of the non-additive terms and will therefore not be able to reproduce experimental gas-phase properties, such as second virial coefficients. Relatively little is known about the non-additive effects. An accurate pair potential for argon (see Section 2.2) only needed the addition of the three-body dispersion term to account for the liquid- and solid-state properties. However, many non-additive contributions to the intermolecular energy are identically zero for neutral spheres like argon. Hence the magnitude and convergence of the non-additive terms could differ considerably between types of molecule and symmetries of packing, as suggested by the different additivity properties of the various contributions to the intermolecular forces (Sections 2.3 and 2.4).

2.2 Calculating the Intermolecular Pair Potential

The intermolecular pair potential of two ground-state, closed-shell, rigid molecules is simply defined as the difference between the energy of the pair of molecules at a given relative separation and orientation and the energy when they are infinitely separated. In principle, one just needs to solve the Schrödinger equation to obtain the energy of the pair at each orientation on the potential-energy surface, and then subtract the energy of the isolated molecules. Unfortunately, this requires such high-quality wavefunctions, with extended basis sets and a high degree of electron correlation, that such supermolecule calculations are unlikely to be routinely possible for even the smallest organic molecules in the foreseeable future. The problem is compounded by the huge number of calculations that would be required to adequately cover the potential-energy surface, as specifying the geometry of a pair of low-symmetry molecules requires six coordinates. Even then, the numerical points have to be fitted to an analytical functional form for use in simulations.

Ab initio surfaces for $(HF)_2$ [4] and $(N_2)_2$ [5] are not well represented by the isotropic atom–atom potential form [6]. Finding an adequate anisotropic atom–atom representation requires guidance from the theory of intermolecular forces, or a prohibitively large number of points. Similarly, theoretical guidance on functional form, and independently determined dispersion coefficients (C_{2n+6}) were required for the derivation of an accurate intermolecular potential for argon (an excellent history of the search for this holy grail is given in [7]). One analytical form, due to Barker et al. [8] for the intermolecular potential of two argon atoms as a function of their separation R,

$$U(R)/\varepsilon = \exp\left[\alpha(1 - R/\rho)\right]\sum_{i=0}^{5} A_i(R/\rho - 1)^i + \sum_{j=0}^{2}\frac{C^*_{2j+6}}{\delta + (R/\rho)^{2j+6}}$$

warns that dozens of parameters are expected to be required for an accurate potential for any polyatomic. So many parameters and a realistic orientation dependence cannot be derived by empirical fitting, even to a wide range of experimental data, because most gas-phase data is determined by averages over the relative orientations of molecules, and solid-state data only samples a limited range of relative orientations. Hence we are reliant on theory or assumption for the functional form of the orientation dependence of the potential, even if it is mainly derived from experimental data or supermolecule calculations.

Our main theoretical insight into intermolecular forces comes from the perturbation theory approach, which decomposes the interaction into different contributions, at different orders of the perturbation theory. The perturbation operator for the interaction energy is the sum of the intermolecular charge–charge interactions $H' = \sum e_i e_j/(4\pi\varepsilon_0 r_{ij})$. When there is no overlap of molecular charge distributions, so that all electron density can be rigorously assigned to one or other molecule, then the intermolecular interaction can be expressed in terms of the properties of the isolated molecules using analytical expressions for the orientation dependence. At shorter range, it is no longer possible to make such an analytical separation, and there are problems associated with electron exchange, antisymmetry and the nature of the zeroth-order Hamiltonian and wavefunction. This had led to a variety of different perturbation-theory treatments that can estimate the various short-range terms starting from the wavefunctions of the isolated molecules. The intermolecular perturbation theory (IMPT) of Hayes and Stone [9] is beginning to be applied to organic molecules, though at such computational cost that only a few points on any intermolecular potential surface can be calculated.

Only the first-order long-range term has been implemented in crystal-structure modelling for organic molecules (see Section 4), so this will be described in more detail. However, it is worth commenting on the other contributions and their likely manifestations in molecular crystals.

2.3 The Intermolecular Potential at Long Range

2.3.1 The Electrostatic Energy

The first-order term in the long-range perturbation theory is the Coulombic interaction between the undistorted molecular charge distributions, that is, the electrostatic energy:

$$U_{\text{e.static}} = \int_{\text{all space}} \frac{\rho^A(\mathbf{r}_1)\rho^B(\mathbf{r}_2)}{|\mathbf{r}_1 - \mathbf{r}_2|}\, \mathrm{d}^3\mathbf{r}_1\, \mathrm{d}^3\mathbf{r}_2 = \langle 0^A 0^B | H' | 0^A 0^B \rangle,$$

where ρ^A is the charge distribution corresponding to the ground-state wavefunction 0^A of molecule A in isolation, and H' is the perturbation operator. Although zero for spherical neutral molecules (atoms), the electrostatic term is significant for most organic molecules. It is the only major contribution to the intermolecular potential that can be either attractive or repulsive. It is strictly pairwise additive. This integral over the molecular charge distributions can be approximated by expanding $1/r_{ij}$, the distance between charges in the different molecules, in terms of a vector **R** between the centres of the two molecules, and the vectors from these centres to the charge in each molecule, and expressing this as a multipole expansion. This converts the expression into an infinite sum over the multipole moments of the isolated molecules, plus terms defining the dependence of the interaction on the relative geometry of the two molecules. The result is the familiar multipole expansion of the electrostatic energy [10], in terms of the charge ($l = 0$), dipole ($l = 1$), quadrupole ($l = 2$), octupole ($l = 3$), hexadecapole ($l = 4$), etc., tensors of the isolated molecules: the interaction between multipole l_A on molecule A and l_B on molecule B depends on R^{-n} (where $n = l_A + l_B + 1$), and the relative orientation of the molecular axes systems (used to define the molecular multipoles). The orientation-dependent terms can be calculated from scalar products of the unit vectors along the two sets of molecular axes and the unit intermolecular vector.

This central multipole series, using the total multipole moments of the molecules, can be used to estimate the electrostatic interaction only at very long range, where only the first non-zero term is significant, but cannot be used in molecular crystal-structure modelling. The first reason is that the central multipole series is very slowly convergent. A central multipole series for, say, nitrobenzene, would have to contain a vast number of terms of the spherical harmonic expansion of the molecular shape before it included even an approximate shape for the nitro group, let alone its detailed charge distribution. Even for diatomics the first term of the central multipole expansion is poor – the lowest-energy crystal-packing for X_2 can depend on whether the same central quadrupole moment is modelled by atomic dipoles or quadrupoles [11]. The central multipole expansion is also not valid in many molecular crystal structures, as demonstrated empirically for pyrazine [12], because, in the solid state, there is overlap of the spheres that contain all the molecular electron density being represented by each multipole series.

The traditional method of calculating the electrostatic energy for organic molecules is to approximate the molecular charge distribution $\rho(r)$ by a set of partial charges at the atomic sites, i.e. as a superposition of spherical atomic charge distributions. Considerable effort has been expended on devising methods of obtaining the best atomic charges. Although Mulliken population charges have been widely used, and are still used by some modelling packages, these are notoriously basis-set-dependent and were not devised for calculating intermolecular forces. The best method of reproducing the electrostatic interactions of a given *ab initio* charge distribution by atomic charges is to fit the charges to reproduce the electrostatic potential around the molecule, as calculated directly by integration over the wavefunction [13]. However, as pointed out in a thorough comparison of these methods [14], atomic charge schemes are incapable of exactly representing the electrostatic potential in a large region around a wide variety of molecules. The errors involved are very dependent on the bonding within the functional groups, their arrangement within the molecule, as well as the specific region around the molecule being sampled.

The limitations on the accuracy of an atomic point charge model come from its fundamental assumption, that the molecule can be accurately represented by a superposition of spherical atomic charge densities. There would be no electrostatic potential outside a molecule that was a superposition of spherical neutral atomic charge densities. Hence, the electrostatic forces arise from the distribution of the valence electrons (strictly from the difference between the molecular charge density and a superposition of neutral spherical atomic densities). The atomic charge model implies that on bonding, charge is redistributed between the (spherical) atoms so that the atomic densities remain spherical, which is a gross misrepresentation of molecular bonding theory. The valence electron density will often include markedly non-spherical features, such as lone pair and π electron density, as well as bonding electron density.

This argument qualitatively suggests that a better model for the molecular charge distribution will be a combination of both the traditional approaches, namely the representation of the charge distribution of each atom by a multipole series. Density-difference maps are very suggestive of atomic multipole series [15], and X-ray charge-density techniques also point to this being a 'natural' method. The only problem with quantifying such a description is the need to divide the molecular charge density into atoms. Many different methods have been proposed [16], including the use of Bader's atoms-in-molecules partitioning [17]. Whilst this is the most fundamental and rigorous method, it is expensive to perform accurately the numerical integrations involved in the partitioning and multipole evaluation, and the sharp edges of the atomic shapes suggest that the multipole series will not converge particularly well. An alternative source of atomic multipole moments is low-temperature X-ray charge-density analysis [18], but these charge distributions are affected by the crystal environment and so the lattice energy evaluated from these moments may be contaminated by induction effects.

A more practical method, designed for optimum convergence and ease of computer analysis, is Stone's distributed multipole analysis (DMA) [19]. This is based on the density matrix of the molecule ρ_{ij}, which expresses the molecular electron density in terms of the Gaussian primitives η that comprise the atomic orbital basis set:

$$\rho(\mathbf{r}) = \sum_{ij} \rho_{ij}\eta_i\eta_j$$

Each contribution to the charge density can be exactly represented (outside itself) by a series of point multipoles at a point determined by the origin and exponents of the two Gaussian primitives involved. If both orbitals are *s* orbitals, then the contribution is represented by a charge, if two *p* orbitals, then a charge, dipole and quadrupole at the overlap centre (the series terminates at the lth multipole, where l is the sum of the angular momenta of the basis functions involved). Thus, for example, the π electron density in aromatic compounds is represented by out-of-plane quadrupole moments ($l = 2$). If both Gaussian functions are on the same atom, then the analysis generates the corresponding limited multipole series on that atom. If they are on different atoms, then the analysis 'moves' the contribution from the overlap centre to the nearest atom, using the relationship between a point multipole at one site to an infinite series at another point. This series converges rapidly provided the distance moved is not too great. Thus for a molecule that is well described by an *sp* basis set, the atomic charges, dipoles and quadrupoles are fundamental to the description of the valence density, whereas the higher moments arise from the displacement of the overlap density and improve the quantitative accuracy of the description.

All the partitioning methods should be able to describe the electrostatic potential outside the molecular charge distribution to essentially the accuracy of the wavefunction, provided that sufficient expansion sites (some schemes have extra sites in addition to the atoms) and orders of multipole are included. Thus, Spackman [20] found that several different distributed multipolar expansions for $(HF)_2$ predicted the electrostatic energies in excellent agreement provided that terms up to quadrupole were included. Electrostatic interaction energies are evaluated using the multipole expansion formulae [21] for each intermolecular pair of sites. A multipole series should strictly be truncated at a given power of the inter-site separation R, rather than at a given order of multipole. Hence, the series should include all terms up to R^{-5} to account for the important quadrupole–quadrupole interactions, and thus will also include the hexadecapole–charge and octupole–dipole terms.

In order to give a theoretically rigorous description of the molecular charge distribution for calculating electrostatic energies, the distributed multipole expansion involves a large number of parameters. An atom, in a site of no symmetry, should have its charge distribution described by a charge, three components of the dipole, five components of quadrupole, seven of octupole and nine of hexadecapole; thus a total of 25 parameters are required to include all terms up to R^{-5} in the expansion. In practice, molecular symmetry, or approximate local symmetry, will considerably reduce the number of significant multipoles, and, for a given molecule, the description could be empirically simplified by testing the effect of omitting the smaller terms on the predicted electrostatic interaction energies.

The main limitations to the accuracy of a distributed multipole description of the electrostatic energy are the neglect of the effects of interpenetration of the charge distributions (see Section 2.4.3) and the quality of the wavefunction used for multipoles. Both basis set and the form (or omission) of electron correlation do affect the charge density: for example, a reasonable basis set (e.g. 6-31G**) SCF wavefunction generally overestimates the total molecular dipole moment by

12–15% [22] and the 0.9 fudge factor commonly used to allow for the effect of electron correlation on the estimated electrostatic energies is based [23] on an average of only eight small polyatomic molecules. A comparison of the electrostatic potential on a grid of points outside N-acetylalanine N′-methylamide, as calculated from the DMAs of various wavefunctions differing in basis set and inclusion of electron correlation [24], suggests that the approximations in the wavefunction, for split valence or better basis sets, will primarily produce a scaling of the electrostatic interactions, rather than any significant change in the orientation dependence of the electrostatic interaction. However, the uncertainty involved in extrapolating this result to other molecules is rapidly disappearing as computer power allows us to increase the size of molecules for which high-quality wavefunctions can be obtained.

Thus, the electrostatic energy, the first-order term in the perturbation theory expansion of the intermolecular potential in the absence of overlap, can be calculated to a high degree of accuracy quite readily for organic molecules. The accuracy is affected by the quality of the wavefunction and the combination of order of multipoles and number of sites, but this error can be estimated.

2.3.2 The Polarization Energy

Second-order perturbation theory provides expressions for the polarization or induction energy. This is the attractive energy term arising from the distortions of the charge density of each molecule due to the field arising from the other (undistorted) molecule:

$$U^{\mathrm{A}}_{\mathrm{induct}} = \sum_{n^{\mathrm{A}} \neq 0^{\mathrm{A}}} \frac{|\langle 0^{\mathrm{A}} 0^{\mathrm{B}} | H' | n^{\mathrm{A}} 0^{\mathrm{B}} \rangle|^2}{E^{\mathrm{A}}_n - E^{\mathrm{A}}_0}$$

The given term represents the additional energy from the changes in the charge density of A, (a molecule whose excited-state wavefunctions n^{A} are of energy E^{A}_n) caused by the presence of B, interacting with the undistorted charge distribution of B. There is a corresponding term that likewise represents the induction energy for the first-order changes in the charge density of B induced by A. This energy is always attractive, as the distortions only occur to lower the energy of the pair.

The application of the multipole expansion allows the induction energy of B to be expressed in terms of the permanent multipole moments of B and the polarizabilities of A, plus the orientation-dependent functions. For all molecules, the first polarizability is the dipole–dipole polarizability, which describes the ease of changing the total dipole moment of the molecule. Thus, for any atom or molecule in the presence of a charge, the first term in the induction energy goes as R^{-4}. However, if another charge is symmetrically placed on the opposite side of the molecule, then there is no induced dipole by symmetry, and the leading induction-energy term is determined by the quadrupolar polarizability. This illustrates that the induction energy is fundamentally non-additive. The induction energy is very important for the intermolecular interactions of ions, but the leading term for the induction energy arising from the presence of a dipolar molecule goes as R^{-6} and so will often be very small for ensembles of neutral molecules, and zero for neutral atoms.

Accurate evaluation of polarization energies is an area of active research and debate for polyatomic molecules. Reliable values for dipolar polarizabilities are difficult to obtain both experimentally and theoretically. Higher polarizabilities are even more sensitive to the quality of the *ab initio* calculation. There are questions of convergence of the series, both at the second order of perturbation theory, and higher orders involving the iteration of the induced moments on both molecules to convergence. For polyatomic molecules in van der Waals contact, the polarizability must be distributed (see distributed multipoles, Section 2.3.1) in order to get a realistic description, with the induced atomic moments formally contributing to higher-order total induced moments.

There are a variety of methods of distributing the polarizability amongst many sites within a molecule. One of the most rigorous is the distributed polarizability analysis of Stone [25]. This provides a very elaborate description of the polarizability, and includes charge–charge polarizabilities, to describe the flow of charge between the different subdivided regions of the molecule under the influence of an external electrostatic field. This is a real effect that will vary in significance according to the molecular bonding, and is likely to be most significant within conjugated systems. The full distributed polarizability model is too elaborate to be used for simulations, and so methods of transforming the non-local polarizabilities involving two sites into only local polarizabilities have been explored [26]. Tests of these methods on saturated and conjugated hydrocarbons show that a fully localized model lacks transferability and accuracy, and a scheme that retains some charge-flow polarizabilities, along with atomic dipolar polarizabilities, is more promising.

The practical difficulties in implementing a necessarily distributed polarizability model within a molecular crystal are going to be greater than for van der Waals molecules, because of the non-additivity. For highly symmetrical small molecules and packings, this will certainly reduce the importance of the polarization effects. However, the local polarization involved in hydrogen bonds, for example, may not be significantly less important than in the van der Waals dimers. Hopefully, accurate polarizability models will be extended to molecular crystals in the near future, giving a first chance to understand the effect of the molecular crystal environment on the molecular charge density. Comparisons of experimental molecular dipole moments in the gas phase and in crystal structures have to be made with care because of the different sources of errors in their determination [27].

2.3.3 The Dispersion Energy

The dispersion energy also appears in second-order perturbation theory as the universal long-range force of the form:

$$U_{\text{disp}} = \sum_{n^{\text{A}} \neq 0^{\text{A}}, n^{\text{B}} \neq 0^{\text{B}}} \frac{|\langle 0^{\text{A}} 0^{\text{B}} | H' | n^{\text{A}} n^{\text{B}} \rangle|^2}{E_n^{\text{A}} - E_0^{\text{A}} + E_n^{\text{B}} - E_0^{\text{B}}}$$

Application of the central multipole expansion produces the usual series,

$$U_{\text{disp}} = -\frac{C_6}{R^6} - \frac{C_8}{R^8} - \frac{C_{10}}{R^{10}} - \dots$$

with odd powers, starting at R^{-7}, appearing for lower-symmetry molecules. The C_n coefficients can be obtained from the charge distributions of the isolated molecules, though this involves integrals over the products of the polarizabilities of the two molecules at imaginary frequencies. The R^{-6} term thus arises from the correlation in the instantaneous dipolar fluctuations in the charge density of the two molecules.

The dispersion energy is the universal attractive glue that leads to the formation of condensed phases. It is additive at second order in perturbation theory, and the form of the three-body term that arises at third order, the Axilrod–Teller term, is also well known from perturbation theory [28]. The Axilrod–Teller term was the only addition to the pair potential for argon that was required to quantitatively account for its solid- and liquid-state properties; this fact may be grounds for optimism that other non-additive dispersion terms are negligible. Whether this will extend to less symmetrical molecules and environments remains to be seen.

Dispersion coefficients can be obtained experimentally in favourable cases from experimental oscillator strengths, but for most molecules one must rely on *ab initio* calculations. Accurate values, particularly for the higher-order dispersion coefficients, require very high-quality wavefunctions, with the high-angular-momentum basis functions being required to reliably describe the susceptibility to lower-symmetry distortions. Again, the central expansion of the dispersion energy is poor for all but the most spherical molecules, and the dispersion coefficients should be distributed over many sites in the molecule. This, in principle, produces dispersion coefficients involving charge-flow between sites and a quadruple sum over sites. However, for the simple cases of nitrogen, chlorine and acetylene, a localization [29] of the polarizabilities to the N, Cl or C atoms plus a central site gave a straightforward site–site sum similar to the familiar atom–atom form, with little loss of accuracy. Convergence is significantly worse if only atomic sites are used, but this may be adequate for larger molecules. Thus the rigorous theory may provide a route to estimating atom–atom dispersion coefficients from molecular wavefunctions in the foreseeable future.

2.4 Contributions to the Intermolecular Potential from Molecular Overlap

When the electron density of molecules overlaps, it is no longer possible to rigorously assign it to one molecule or another. The wavefunction for the 'supermolecule' has to be antisymmetric with respect to the exchange of any pair of electrons, and the theory is the same as for intramolecular bonding. This means that there are no analytical theories for the functional forms of these short-range contributions, nor can their strengths be rigorously related to the properties of the isolated molecules. However, perturbation theories do provide guidance on the qualitative nature of the interaction, and can be used to numerically quantify the various contributions.

2.4.1 Exchange-Repulsion

The new first-order terms arising from the overlap of the charge distributions result in net intermolecular repulsion. The overlapping charge densities cannot occupy the same space (Pauli exclusion principle), which produces repulsive forces that dominate the attractive forces arising from the electron exchange between the molecules. The two terms are usually taken together, and for atoms can be represented as $f(R)\exp(-\alpha R)$, where $f(R)$ is a slowly varying polynomial function of the separation R. It is approximately additive, at the level of overlap normally found in molecular crystals and van der Waals complexes.

The exchange-repulsion can be evaluated numerically from supermolecule calculations at the SCF level, or perturbation-theory calculations, and fitted by model potentials. Although the radial dependence of the usual isotropic atom–atom exponential repulsion could be improved, the main error appears to come from the assumption that the repulsive wall around each atom is spherical. The non-spherical features in the molecular charge density that give rise to the atomic multipoles also affect the repulsion interaction, leading to anisotropy in the atom–atom repulsive potential. The functional form of the anisotropy will depend on the orientation of the lone pairs, π electrons and other features. For example, the fit to a $(N_2)_2$ [6] *ab initio* potential-energy surface with the atom–atom potential

$$U = \sum_{i\in A}\sum_{j\in B} \varepsilon(\alpha\rho_0 - 6)^{-1}[6\exp(-\alpha(R_{ij} - \rho(\Omega_{ij}))) - \alpha\rho_0^7/(R_{ij} - \rho(\Omega_{ij}) + \rho_0)^6] + U_{\text{electrostatic}}$$

was found to be far superior when the shape of the atoms was included by making the minimum energy separation ρ depend on the orientation of the two atoms, than when it was isotropic ($\rho(\Omega) = \rho_0$). Almost a five-fold reduction in the r.m.s. error in the representation of the *ab initio* SCF surface points was obtained with the form

$$\rho(\Omega) = \rho_0 + \rho_1(\mathbf{z}_1\cdot\mathbf{R} - \mathbf{z}_2\cdot\mathbf{R}) + \rho_2\tfrac{1}{2}(3\mathbf{z}_1\cdot\mathbf{R}^2 + 3\mathbf{z}_2\cdot\mathbf{R}^2 - 2)$$

where the orientation dependence is expressed in terms of scalar products between the unit inter-site vector $\mathbf{R}$ and unit bond vectors $\mathbf{z}_1$ and $\mathbf{z}_2$, defined to point from the centre of the molecule towards the atoms whose interactions are being calculated. The combination of anisotropy of these two terms (which are formed from the first two Legendre polynomials) shifts the repulsive wall out at the end of the molecule and close to the intramolecular bond, allowing closer approach in the plane through the atom perpendicular to the intramolecular bond, compatible with the charge distribution in N_2.

Anisotropic repulsion potentials have not yet been widely used for modeling molecular crystals, though the use of a formalism that can model both the nitrogen lone pair and the C–H bond, to provide six-site potentials for the azabenzenes [30] or carbon-only site potentials for the saturated hydrocarbons [31], has computational advantages as well as the flexibility to be refined as more information about the anisotropy of the repulsive wall becomes available. The anisotropy of the repulsive wall is important in determining the van der Waals contact distance in a crystal, and this affects the lattice energy mainly through the long-range contributions. Certainly, plausible variations in atomic shape (as represented by anisotropic atom–atom potentials) can lead to significant changes in the relative stability of the crystal structures of X_2 molecules [11].

One approximate method of determining the anisotropic form and parameters for the exchange-repulsion interaction is based on the assumption that it is proportional to the overlap of the undistorted charge distributions [32]. This provides analytical forms where only the proportionality constant has to be determined. This method has been justified by comparisons for N_2, F_2 and Cl_2 against supermolecule SCF surfaces. So, once again, the form of this contribution to the intermolecular forces can be derived, to a large extent, from the charge densities of the isolated molecules.

2.4.2 Charge-Transfer

A further term that arises at second order in perturbation theory is the transfer of charge from the occupied orbitals of one molecule to the unoccupied orbitals of the other. These charge-transfer or intermolecular bonding effects can be evaluated by perturbation theory [33], or Morokuma decomposition of SCF supermolecule calculations [34], and have been found to have an approximately exponential decay with separation. The values of the charge-transfer contribution will depend on the definition used to separate it from the exchange-repulsion and the long-range polarization [33]. However, a far more important problem is that its magnitude can be grossly overestimated because of contamination with basis-set superposition error. The apparent use of the orbitals based on one molecule to describe the charge distribution of the other can include the use of these orbitals to improve the description of the intramolecular interactions within one molecule (i.e. give a spurious energy-lowering because of the inadequate basis sets of the isolated molecules) as well as describe the real transfer of charge between the molecules. This problem has been overcome recently, through better theoretical methods [33] and better basis sets, but early estimates of charge-transfer effects must be viewed with considerable caution.

The distinction between charge-transfer (sometimes loosely termed 'weak intermolecular bonding' or 'specific attractive forces') and exchange-repulsion has been a matter of some controversy in the understanding of molecular crystals. The marked anisotropy in the effective van der Waals radii of some atoms [35], particularly halogens, has been attributed to weak intermolecular bonding providing a specific attraction in the directions that allow interactions of the lone-pair density. The original controversy centred on the layered herring-bone crystal structure of Cl_2, which certainly cannot be accounted for by an isotropic atom–atom potential plus electrostatic model. The addition of weak intermolecular bonds (represented by a Morse potential) certainly stabilizes the observed structure [36], but so do empirical [37], and more recently *ab initio*-based [38], anisotropic repulsion potentials. The anisotropic repulsion model is supported by an electron-density difference map [39], which shows the lone-pair electron density from one molecule fitting into the depression in the repulsive wall at the end of the intramolecular bond, giving the shorter van der Waals contact distance. The more general proposal that the frequent occurrence of short intermolecular contacts in chlorine–chlorine interactions in molecular crystal structures is evidence for a significant specific attractive force [40], has recently been countered [41] by intermolecular-perturbation theory calculations, electrostatic modeling and a reanalysis of the database evidence. The charge-transfer term is very small, and the anisotropy of the exchange-repulsion and electrostatic terms are sufficient to account for the observed anisotropy of the Cl···Cl interactions, which can be mildly attractive for some van der Waals contact orientations.

The debate as to whether charge-transfer is predominantly responsible for the observed anisotropy in some van der Waals contacts is vital for the design of model intermolecular potentials capable of predicting the molecular crystal-packing. The charge-transfer term is very non-additive, and it would be difficult to implement a physically and theoretically justified model for this contribution in a crystal-structure simulation. The way forward for accurate potentials involving anisotropic atom–atom repulsions is much clearer (Section 2.4.1) and it can be treated as pairwise additive. Nevertheless, both effects are genuine, and their relative importance will depend on the molecular system involved.

2.4.3 Damping of Long-range Terms

A final effect that occurs when the molecular charge distributions overlap is that this modifies the long-range electrostatic, induction and dispersion contributions, damping down the R^{-n} separation dependence and thus avoiding the singularity as $R \rightarrow 0$. These effects are not negligible in the van der Waals region. The penetration energy, defined as the difference between the first-order electrostatic energy and its converged (distributed) point multipole expansion, has been estimated [42] to contribute 20% of the total electrostatic energy of the formamide/formaldehyde hydrogen bond.

The damping terms are usually absorbed into any empirical short-range exponential repulsive potential. Their form and strength, a subject of current research for atomic intermolecular interactions [43], is therefore important when a potential is being constructed by summing up models for each of the other contributions. Work towards an accurate intermolecular potential for $(N_2)_2$ [44], showed that the assumed damping functions can have a significant effect on the intermolecular potential in the van der Waals region. Hence, such terms have an effect on solid-state properties.

3. CONSTRUCTION OF ACCURATE INTERMOLECULAR POTENTIALS FOR ORGANIC MOLECULES

3.1 Existing Systematic Potentials

The above analysis suggests that quite accurate intermolecular potentials can be constructed by summing anisotropic atom–atom-type models for the different contributions to the potential. If these terms are mainly parameterized from the molecular charge density (i.e. reliant on one high-quality wavefunction for the isolated molecule), rather than from sufficient calculations on a pair of molecules to define the potential-energy surface, then the methods should, with time, be useful for constructing accurate intermolecular potentials for organic molecules and eventually bio-molecules.

Considerable efforts are being made with this approach, varying in the level of approximation according to both the molecular system and the accuracy required for simulating the properties of interest. Perhaps the best example, with a careful and full parameterization of the electrostatic, induction, dispersion and short-range repulsion energies is the water potential of Millot and Stone [45]. Its complexity suggests that potentials for simulations may in future be derived by constructing the theoretically best possible potential and then systematically simplifying it by omitting terms that have a negligible effect on those regions of the potential-energy surface that are important for the simulation. Another systematic approach [38] gave a Cl_2 potential that was able to account for a wide range of properties of the solid and liquid state, including the structure (Section 2.4.2), with only a few repulsion parameters being determined empirically. Other related schemes have been used for the urea–water system [46], and for ligands binding to Zn^{2+} [47].

Thus, considerable progress is being made towards accurate intermolecular potentials for small rigid polyatomics. Whilst a 'recipe' is still emerging in terms of a synthetic potential approach, it is clear that anisotropy is required in the atom–atom intermolecular potential model, and that the form of the potential will be determined by the form of the valence charge distribution.

3.2 What are the Important Terms in the Potential for Modelling the Crystal Structures of Polar Organic Molecules?

In order to improve on the isotropic atom–atom 6-exp potential in modelling molecular crystal structures, the first stage is to consider whether an improvement to just one aspect of the intermolecular potential would be worthwhile. Molecular crystal structures are very sensitive to the intermolecular potential in the region corresponding to van der Waals contact. The orientation dependence of the potential is particularly important, as many lattice energy minimizations are dominated by molecules rotating or sliding relative to each other.

A key pointer emerges from the modelling of van der Waals complexes, which are also sensitive to the same aspects of the potential, differing from molecular crystal structures in that the forces between a pair of molecules, rather than a packed assembly, are optimized. Buckingham and Fowler [48] demonstrated that the structure of over two dozen complexes of small polyatomics, mainly involving hydrogen bonds, could be predicted by optimizing the electrostatic interaction of the molecules, within the region allowed by a hard-sphere model of the effective molecular shapes. This result depended on using a realistic, distributed multipole model for the electrostatic interaction. Since these structures generally obeyed the Legon–Millen rule [49], that the hydrogen bond forms to the lone-pair or π electron density in the acceptor, it is not surprising that their prediction required a sufficiently detailed electrostatic model to include the anisotropic electrostatic forces arising from the lone-pair and π electron densities.

The dominance of electrostatic forces in determining the orientation dependence of the potential-energy surfaces of van der Waals complexes, implicit in these results, was confirmed by intermolecular-perturbation theory calculations [50] on some of the complexes. These showed that the orientation dependence of the electrostatic energy (calculated from a DMA) did indeed approximately parallel that of the total potential, because there was considerable cancellation in the orientation dependence of the other contributions. The few failures of the Buckingham–Fowler model could be attributed to either a very flat electrostatic (and total) potential surface (e.g. $H_2O\cdots HC{\equiv}CH$), or where the anisotropy of the repulsion was sufficiently marked that the separations implicit in the hard-sphere model produced the wrong relative energies for two possible structure (e.g. F—Cl···F—H is more stable than Cl—F···HF). Thus, in all but some particularly sensitive cases, the use of a realistic electrostatic model with a crude isotropic representation of the other contributions gives a semi-quantitatively correct model potential for studying van der Waals molecules. This model is not easily improved without a full-scale determination of the strength and anisotropy of all terms [51].

A distributed multipole electrostatic plus an isotropic atom–atom repulsion–dispersion model cannot be so rigorously tested for organic molecules. However, it does appear to give a reasonable approximation to the orientation dependence of the intermolecular potential, in that such models have been used successfully for modelling van der Waals complexes of aromatic molecules [52], nucleic base interactions [53], and the preferred relative orientations of protein side-chains [54]. It certainly appears to give a reasonable approximation to the orientation dependence of the N—H···O═C hydrogen bond [55], as shown by intermolecular-perturbation theory calculations on the formamide/formaldehyde model system. Hence, it seems a reasonable assumption on which to try to improve the modelling of organic intermolecular interactions in crystal structures.

3.3 Transferability in Organic Model Intermolecular Potentials

A key question that arises in modelling organic crystals is the degree of transferability of atom–atom model potentials between molecules and between conformations. If an atomic charge distribution differs between the two molecules, then all its intermolecular interactions will also differ, but the change will be less important for some contributions than others. The electrostatic term cannot be as transferable as the repulsion and dispersion coefficients are usually assumed to be: for example, the charge q for hydrogen and $-q$ for carbon which are required to model benzene cannot be used for hexane as this would give the molecule a net charge of $-8q$. Nevertheless, the identifiable properties of specific functional groups, which make organic chemistry so systematic, implies that there is some transferability of functional-group charge distributions between molecules. An analysis of the transferability of distributed atomic multipoles, as calculated from the 3-21G SCF wavefunctions of a range of amides and small peptides, showed [56] that the multipoles were transferable, to a reasonable approximation, provided the neighbouring functional groups were the same. However, atomic charges were very sensitive to the nature of the bonded atoms; for example, a primary amide nitrogen (CO**N**H_2) had an average charge of −0.805 (±0.015)e, whereas a secondary amide nitrogen where one hydrogen had been replaced by a methyl group ($CONHCH_3$) had an average charge of −0.547 (±0.016)e, and further methyl substitution to $CON(CH_3)_2$ reduced the average nitrogen charge to −0.289 (±0.018)e. The higher anisotropic multipole moments were also similarly transferable, when referred to equivalent local axes. All the multipole moments vary even more between different hybridization states. The cautionary concept that the charge distribution is only approximately transferable between atoms subject to similar short-range inductive effects (i.e. having the same bonded functional groups), is also important when building up molecular electron densities from fragment distributions [57].

The transferability of molecular charge distribution fragments is essential for building up distributed multipole models of large bio-molecules. Tests on small polypeptides suggest that this can be done reasonably accurately [58]. Assuming that the charge distribution is independent of conformation produces rather more significant errors in the predicted electrostatic interactions, using either potential-derived charges [59] or atomic multipoles [60]. It seems likely that this error is mainly due to a physical effect, related to the genuine changes that occur in the charge distribution with conformation. Although long-range through-space effects can be modelled using distributed polarizability models [58], the through-bond effects, arising from changes in conjugation for example, are more serious. Such problems may be overcome by expanding each multipole term as a Fourier series in the torsion angle [61]. Thus, the use of distributed multipole electrostatic models for large organic systems is perfectly feasible, though some care may be needed when studying conformationally flexible molecules.

4. DISTRIBUTED MULTIPOLE MODELS APPLIED TO MOLECULAR CRYSTAL-STRUCTURE MODELLING

The above discussion shows that many different intermolecular effects will have to be quantified in order to obtain a definitive model for the forces that determine the properties of molecular crystals. A definitive model is needed for confidence in the predicted values of all properties, such as lattice energies, phonon modes, thermal expansion coefficients, etc. However, when just aiming at a model potential that is usually good enough to determine the crystal structure, the evidence from the structures of van der Waals molecules suggests that the addition of an accurate, distributed multipole electrostatic model could be worthwhile.

There are varied opinions on the importance of electrostatic forces in determining crystal structures. For example, Perlstein [62] found that the electrostatic forces were relatively unimportant in determining the structures of a range of large semi-flexible organic molecules, whereas Williams and Weller [63] found it necessary to add additional lone-pair sites to obtain a sufficiently realistic representation of the electrostatic forces to predict the crystal structures of the azabenzenes. Similarly, the lowest-energy structure of a heteratomic diatomic can depend on whether the same total quadrupole is modelled by atomic dipoles or quadrupoles [11]. These examples emphasize that the importance of electrostatic forces depends not only on the electronegativity of the atoms involved, but also on the overall shape of the molecule. Close-packing knobbly molecules may form tightly bound 3D jigsaws, whereas planar or disc- or cylinder-shaped molecules can be effectively close-packed in a wide range of structures, and be able to slide and rotate relative to each other relatively easily without encountering large barriers from the repulsive potential. Such structures will be far more sensitive to the orientation dependence of the potential, and therefore the electrostatic forces, than structures with a rigid network of protruding atoms fitting into hollows.

4.1 Computer Programs Using Distributed Multipole Models

The main barrier to going beyond the isotropic site–site intermolecular potential model for simulating molecular crystal structures was the need to adapt a computer program to use distributed multipole models. This was not trivial because the force between two sites is no longer central, and torques are generated. For example, an atomic dipole, arising from lone-pair density, has its orientation fixed relative to the intramolecular bonds to that atom. Hence the electrostatic interaction of that dipole with a dipole in another molecule will depend on their relative orientation (defined relative to the local axes), producing equal and opposite forces on the two dipoles, usually not parallel to the inter-dipole vector. The two dipoles will also tend to align anti-parallel, thus producing a torque acting to rotate each molecule about the dipolar atom.

The movement resulting from intermolecular electrostatic forces and torques acting between a given pair of atoms will be considerably modified by the intramolecular forces as well as the other intermolecular atom–atom forces. Thus, it is advantageous both computationally and to avoid having to parameterize intramolecular forces, to treat each molecule as rigid. (This could easily be extended to treat a molecule as comprised of rigid fragments that are joined by an intramolecular potential, if necessary.) The rigid molecule then rotates and translates under the net force and torque acting at its centre of mass and arising from all the intermolecular atom–atom interactions. The torques arising from the anisotropic multipoles can be summed to give a total torque about the centre of mass of each rigid molecule, which is in addition to the net torque created by resolving the atom–atom forces into a total force and torque at the centre of the molecule (see a more rigorous argument in [64]).

Spherical tensor theory can be used to derive the orientation dependence of all the terms in the multipole series up to R^{-5}, in terms of the scalar products of the local (molecular) axis vectors, relative to which the multipoles are defined, and the inter-site vector. Various recursive relationships make such expressions easy to derive, program and evaluate. Nevertheless, only relatively recently have the analytical expressions for the corresponding forces, torques and second derivatives with respect to molecular translation and orientation been evaluated [65]. For molecular crystal structure simulations, the derivatives with respect to the changes in unit cell parameters have also been derived [64] and implemented.

The program DMAREL [64] has been written to simulate static crystals of molecules whose electrostatic interactions are represented by a distributed multipole model, including all terms to R^{-5}. The program uses a modified Newton–Raphson procedure, which makes occasional use of some analytical second derivatives, to find a stable crystal structure by reducing to zero the forces and torques on the molecules, and the lattice strain (within a preset tolerance). The starting-point crystal structure, usually taken as the experimental crystal structure as produced by an FDAT file from the Cambridge Structural Database [66], is converted into a cartesian representation. Thus, the lattice vectors, and positions and orientations of each molecule in the unit cell are optimized independently, so the crystal symmetry is not enforced. However, the symmetry of the force field in the starting structure may prevent symmetry-breaking, so that the fact that symmetry is usually maintained does not rule out the possibility that global minima could occur for a less symmetrical structure.

The use of distributed multipoles in crystal structure simulations also requires the evaluation of the contributions to the lattice energy from the anisotropic long-range multipole interactions. Direct summation is far more convergent when applied so that entire neutral molecules are included in the sum, with the cut-off radius referring to centre of mass separation, rather than to atom–atom distance. Nevertheless, the charge–charge, charge–dipole and dipole–dipole terms die away individually as R^{-1}, R^{-2} and R^{-3} respectively, and although there is some cancellation between the contributions from whole molecules, it was found advantageous to develop the Ewald-summation technique, as routinely applied to charge–charge interaction, to cover these anisotropic contributions also. These summation techniques could be extended to higher terms, but become more complex and less worthwhile as the sums converge more quickly in direct space. Empirically it was found that a direct summation convergence limit of 20 Å both on molecular centres for the remaining electrostatic terms and on atomic separations for the repulsion–dispersion terms, was sufficient to be confident of convergence within 1 kJ/mol over a range of polar neutral molecules. A smaller summation limit could be used for some molecules and crystal symmetries, or for purposes when lower accuracy was required.

The computational requirements of a DMAREL minimization depend on the number and size of molecules in the unit cell, the range used for the lattice summations, and, of course, the number of iterations required to find a minimum. The calculations are considerably more time-consuming than isotropic atom–

atom potentials, but well within the range of modern computers. The current version of DMAREL takes about half an hour of CPU use on a R8000 processor machine to minimize crystal structures like those discussed in Section 4.2. The spectacular increase in computer power over the last decade has meant that optimizing computer performance was not a high priority in the development of the program. An alternative approach has been used by Ritchie et al. [67] who have developed an older crystal-structure optimization program (Williams's PCK83) to use anisotropic multipole moments.

4.2 Testing DMA-based Model Potential for Modelling Molecular Crystal Structures

The use of realistic electrostatic models is expected to be most important for polar and hydrogen-bonded molecules, which are systems with non-spherical features such as π electrons and lone-pair density and often occur in molecules whose shapes are unlikely to determine the crystal-packing. This includes a wide range of organic molecules; therefore we have studied molecules ranging from nucleic acid bases and amides to nitrobenzenes and nitroanilines, to cover examples of some simple organic pharmaceutical, non-linear optic and explosive materials. Over forty molecules have been studied to date [68], to sample various combinations of aromatic rings, heterocyclic rings, amine, nitro and amide groups. Thus some of the molecular crystals include N—H···O═C and N—H···N hydrogen bonds and π–π interactions, and several are layered structures. All the molecules tested could be reasonably modelled as rigid, to a first approximation, to avoid the complication of conformational changes. The choice was made, with some degree of arbitrariness, from molecules where a reasonable-quality, approximately room-temperature crystal structure was found in the Cambridge Structural Database.

4.2.1 Methodology Adopted

The molecular structures were taken directly from the experimental crystal structures, with the hydrogen atoms positioned to give a standard C—H bond length of 1.08 Å for C—H and 1.01 Å for N—H along the experimental bond direction [69].

The calculations used an atom–atom model intermolecular potential that comprised an *ab initio*-based distributed multipole model for the electrostatic contribution and an empirically fitted atom–atom 6-exp repulsion–dispersion potential. The DMA electrostatic model was obtained from the SCF wavefunction of the isolated molecule, calculated using a 6-31G** basis set [70], within the program suite CADPAC [71]. The full multipole expansion of the atom–atom electrostatic interaction energy was used, including all terms up to R^{-5} and hence up to 25 independent multipole components to represent the electronic charge distribution around each atom. Thus, the main error in the evaluation of the electrostatic contribution to the lattice energies, excluding the effect of penetration into the molecular charge distributions, was the accuracy of the *ab initio* wavefunction. The lack of electron correlation can be crudely accounted for, by scaling all multipoles by a factor of 0.9 (Section 2.3.1). The sensitivity of the predicted crystal structures to the electrostatic model was tested by considering three of them, the full DMA of the SCF 6-31G** wavefunction (DMA model), its isotropic component (i.e. only the charge terms) to investigate the importance of the anisotropic multipoles (Q model), and the full set of multipoles multiplied by the scaling factor of 0.9 (0.9DMA model). The last model should be the best approximation to the actual electrostatic forces between the molecules.

Given our current inability (Section 2) to derive the other terms in the intermolecular potential from the molecular wavefunctions for organic molecules, the rest of the model potential was kept as simple as possible, so that only the electrostatic part was changed from the traditional approach to modelling molecular crystals. Hence the sum of all other contributions to the intermolecular potential was assumed to have the isotropic atom–atom form

$$U = \sum_{ik} U_{ik} = \sum_{ik} A_{\iota\kappa} \exp(-B_{\iota\kappa} R_{ik}) - C_{\iota\kappa}/R_{ik}^6,$$

where atoms i and k are of types ι and κ respectively. This potential depends only on the separation R_{ik} of the atoms in different molecules, and the types of the atoms concerned. The types considered for this wide-ranging dataset of molecules were carbon (C), nitrogen (N), oxygen (O), hydrogen bonded to carbon (H_c) and hydrogen bonded to nitrogen (H_p), to keep the number of types to a minimum. The distinction between polar H_p and non-polar H_c is necessary to model the smaller effective van der Waals radius of protons involved in hydrogen-bonding. This reflects the major difference in the electron density associated with the two types of hydrogen atoms. By not distinguishing between, for example, nitrogen atoms in amide or nitro groups, we are effectively assuming that the variations in the charge distributions associated with different hybridization and bonding environments of nitrogen, oxygen and carbon atoms need only be represented in the electrostatic model.

The empirical potential repulsion–dispersion parameters were taken from the work of the Williams group, with the parameters for C, H_c and N derived from fitting to azahydrocarbon crystal structures [72], and the O parameters from their fitting to oxohydrocarbons [73]. These potentials were fitted in conjunction with an electrostatic model that used potential-derived charges, including lone-pair charge sites on the azabenzenes. The starting point for the polar hydrogen potentials was taken from the $O\cdots H_p$ potential fitted to intermolecular perturbation theory calculations of the exchange-repulsion, penetration and dispersion interaction between formamide and formaldehyde in the N—H···O═C hydrogen-bonding region [55]. The heteroatomic parameters were fixed using the traditional combining rules:

$$A_{\iota\kappa} = (A_{\iota\iota}A_{\kappa\kappa})^{1/2},\ B_{\iota\kappa} = \tfrac{1}{2}(B_{\iota\iota} + B_{\kappa\kappa}),\ C_{\iota\kappa} = (C_{\iota\iota}C_{\kappa\kappa})^{1/2}.$$

The combining rule for the $C_{\iota\kappa}$ coefficients has some physical justification at the level of the Unsöld approximation to the R^{-6} dispersion coefficient, but the other rules, though widely used, are poorly justified and known to be limited in their accuracy [74]. This composite repulsion–dispersion potential, whilst using one of the most complete sets of parameters developed for molecular crystals in conjunction with any estimate of the electrostatic forces, is nevertheless being used outside the realm for which the individual parameters were derived and tested. However, this initial starting-point potential (denoted EST) proved remarkably successful. The only improvement that could be made to the potential (denoted FIT), despite various attempts, was the refinement of the H_p parameters, to give a slightly deeper well and thereby partially absorb the polarization and charge-transfer contribution to the hydrogen bond energy.

4.2.2 Crystal-Structure Reproduction Using Distributed Multipole Electrostatic Models

The minimum in the lattice energy found by relaxing the structure from the experimental structure was usually remarkably close to the experimental structure. Figure 2.1 gives the r.m.s. percentage error in the independent cell edges for a wide selection of the polar organic molecules containing carbon, hydrogen, nitrogen and sometimes oxygen, as a summary of the published detailed calculated structures [68]. From this graph, it is clear that most of the structures are being reproduced within an error of a few percentage points, and most of the exceptions shown contain NH_2 groups (a few other exceptions are discussed in Section 4.2.3).

At this level of accuracy, the error inherent in using a static minimization method (i.e. a method that neglects molecular vibrations) to model these room-temperature crystal structures is probably important. From the relatively few organic systems where the crystal structure has been determined at a really low temperature, as well as at room temperature, it appears that lattice expansions are generally of the order of a few percentage points, but this can vary considerably between the different cell vectors, with some cell lengths even contracting with increasing temperature [75]. Although thermal effects will be partially absorbed into the empirical repulsion–dispersion parameters, this cannot correctly represent such an anisotropic effect. Hence, at the level of a few percent error in the lattice vectors, the discrepancies from experiment may well result

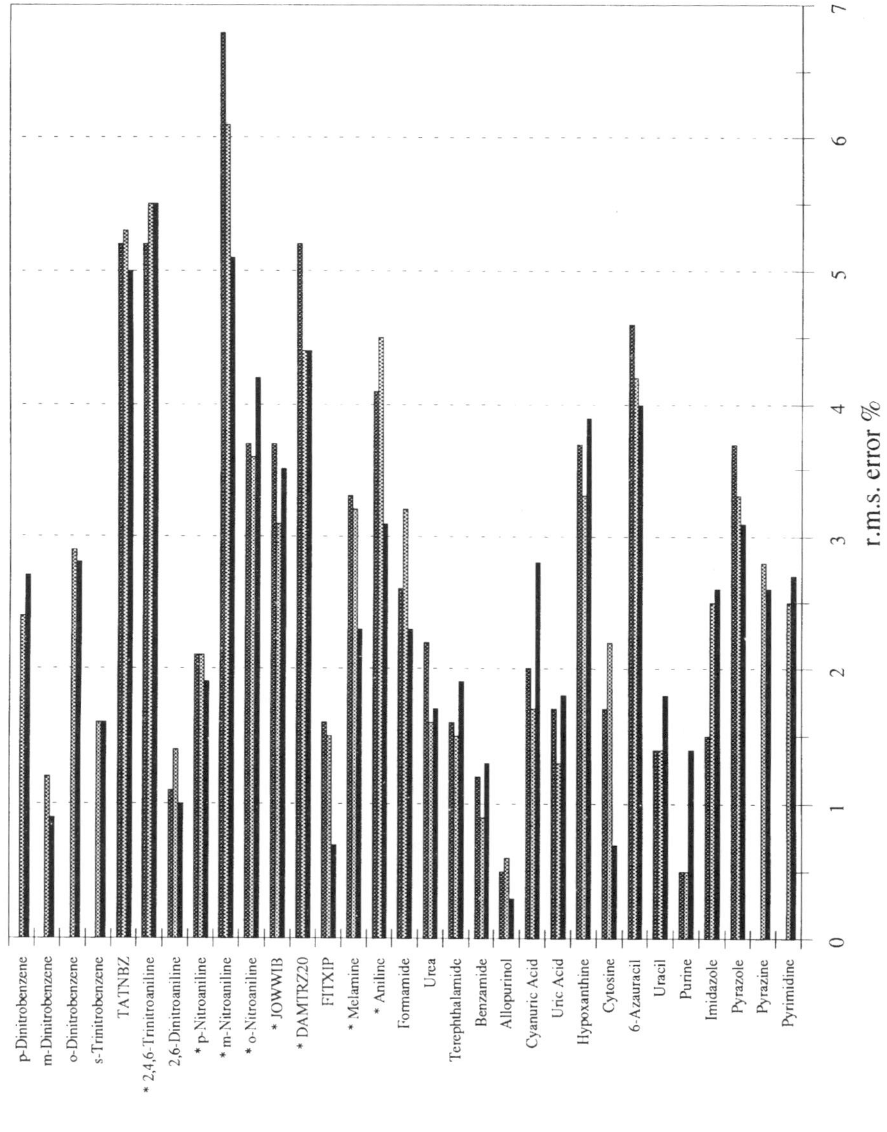
p-Dinitrobenzene
m-Dinitrobenzene
o-Dinitrobenzene
s-Trinitrobenzene
TATNBZ
* 2,4,6-Trinitroaniline
2,6-Dinitroaniline
* p-Nitroaniline
* m-Nitroaniline
* o-Nitroaniline
* JOWWIB
* DAMTRZ20
FITXIP
* Melamine
* Aniline
Formamide
Urea
Terephthalamide
Benzamide
Allopurinol
Cyanuric Acid
Uric Acid
Hypoxanthine
Cytosine
6-Azauracil
Uracil
Purine
Imidazole
Pyrazole
Pyrazine
Pyrimidine
0
1
2
3
4
5
6
7
r.m.s. error %

from the static lattice model, and cannot be ascribed to deficiencies in the intermolecular potential.

The calculated lattice energies are compared with experimental heats of sublimation, when available [76], as in Figure 2.2. The difference is generally within the 3–4 kcal/mol (12–17 kJ/mol) that 'should not cause any concern when judging the quality of the parameters of a potential' [1].

The predicted crystal structures are affected by the scaling of the atomic multipoles to represent the effect of electron correlation, as a comparison of the results for EST + 0.9DMA and EST + DMA in Figure 2.1 demonstrates. Although most structures improve when the relative strengths of the electrostatic and repulsive forces are reduced to allow for the effects of electron correlation, the effect is minor. One exception is indazole, which is not included in Figure 2.1, as the scaling of the electrostatic contribution increases the r.m.s. % error from 5.6% to 9.3%, which is off the scale. This structure is so sensitive to the balance of the electrostatic forces that the error in the crude scaling factor is clearly important in this exceptional case.

The effect of this scaling factor on the predicted lattice energy can be very significant, depending on the proportion of the lattice energy that is electrostatic. This is often a very high proportion of the total for polar and hydrogen-bonded molecules, as shown in Figure 2.2, because strong attractive electrostatic forces pull the molecules together, increasing the repulsive energy so that it cancels out much of the attractive dispersion contribution. The scaling factor of 0.9 applied to the multipoles reduces the electrostatic contribution to the lattice energy by $0.9 \times 0.9 = 0.81$. Thus the uncertainty in the scaling factor can produce a significant error, even compared to the typical experimental errors in heats of

---

Figure 2.1 Differences between the calculated and the experimental room-temperature crystal structure, using distributed multipole models for the electrostatic forces, for a range of organic molecules. The molecules include examples of the nitrobenzenes, nitroanilines, amides, nitrogen-containing heterocycles including nucleic acid bases to purines, and azabenzenes. (The molecules with abbreviated names are: TATNBZ 1,3,5-triamino-2,4,6-trinitrobenzene; JOWWIB 3-amino-5-nitro-1-2-4-triazole; DAMTRZ20 3,5-diamino-1H-1,2,4-triazole; FITXIP 3,6-diamino-1,2,4,5-tetrazine.) * denotes molecules whose NH_2 groups are partially sp^3 hybridized. The error is presented as the r.m.s. % error in the independent cell edges, i.e.

$$\left[\frac{1}{n}\sum_{i=1}^{n}\left(\frac{100(a_{i,\text{calc}} - a_{i,\text{obs}})}{a_{i,\text{obs}}}\right)^2\right]^{1/2}$$

where $n = 3$ except for urea and FITXIP ($n = 2$).

The model potentials are represented by the bars in each group, from left to right: EST + 0.9DMA; EST + DMA; FIT + 0.9DMA

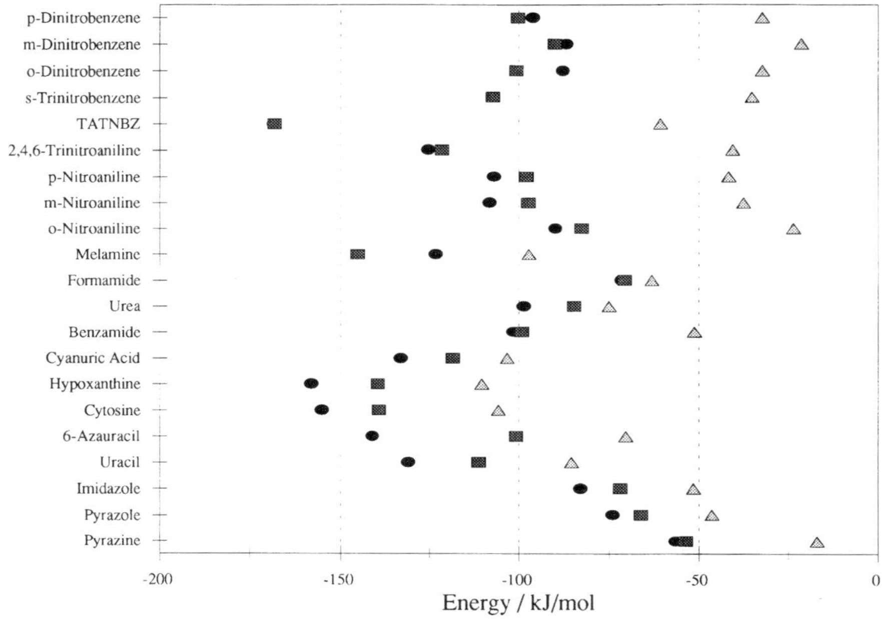

Figure 2.2 Comparison of predicted lattice energies (□), with experimental heats of sublimation ΔH_{sub} (○) where available [76]. The lattice energies were calculated using 0.9DMA + FIT at the lattice energy minimum found by relaxing the crystal structure from the experimental structure. The scaled electrostatic contribution (0.9DMA) at the experimental structure is denoted by (△). See also Chapter 3, Appendix

sublimation. Hence, it would be better to use correlated wavefunctions to obtain DMAs for predicting the lattice energy.

The overall sensitivity of the predicted structures to the electrostatic model can be crudely gauged by removing the anisotropic moments from the electrostatic model. This gave predictions with errors of greater than 4% in all but eight cases, with a comparable number with errors of greater than 10% corresponding to a qualitatively different structure. This demonstrates that these structures are not determined by the molecular shape, and that the anisotropic multipoles are making a significant contribution to stabilizing the observed structure. It does not demonstrate that a more carefully optimized atomic charge model could not model any given crystal structure, although this has already been demonstrated for the azabenzenes [63].

The polar hydrogen interactions are important in determining the hydrogen-bonded crystal structures. Figure 2.1 shows that there is some sensitivity to their repulsion–dispersion parameters. However, it is more significant that most of the worst crystal-structure predictions in Figure 2.1 correspond to molecules where the C—N bond to the NH_2 group is sufficiently long ($\geqslant 1.34$ Å) to suggest a significant degree of sp^3 hybridization, and hence uncertainty in the direction of the N—H bond. Altering the bond direction from experimental to an idealized sp^2 geometry had a variable effect, the most significant being a reduction in the error for *m*-nitroaniline reduced from 5.1% to 1.4%. Thus, uncertainty in the geometry, or dynamic movement, of groups such as NH_2 can also be a problem for static energy minimization.

4.2.3 Molecules that Require More Sophisticated Potentials

Thus, for the vast majority of the molecules so far considered, it would appear that a realistic electrostatic model, plus a simple repulsion–dispersion potential, is a sufficiently good intermolecular potential that the remaining errors in the potential are no more significant than those involved in using static minimization techniques and rigid molecules. There are exceptions, which point to the increased sensitivity of certain molecular packing to the details of the potential. Indazole's sensitivity to the electrostatic scaling is one example. Another is *s*-tetrazine, where the EST + 0.9DMA model gives an error of 6.7%, comparable to that of Williams and Weller (7.3%) [63], and rather larger than the 2.3% obtained when an anisotropic N···N repulsion model was used [30]. The close N···N contacts, which are inevitable in the crystal structure of $C_2H_2N_4$, are probably a rare sensitive probe to the anisotropy of repulsion caused by the nitrogen lone-pair. Similarly, no isotropic chlorine potential could hope to reproduce the anisotropic variation in Cl···Cl van der Waals separation, discussed in Section 2.4.2, as clearly analysed for the polymorphs of *p*-dichlorobenzene [77]. The little electron density associated with the hydrogen atom is clearly not spherically distributed, as recognized by the displacement of the hydrogen site 0.07 Å into the bond in the Williams' potentials [78]. Although we have not used such a displacement in the model described above, it makes a very marked difference to the predicted structure of benzene. Hence, anisotropic repulsion potentials will be necessary for modelling some molecular crystal structures, depending on both the packing and the functional group. Fortunately, such an extension to a crystal-structure prediction program that already handles anisotropic atom–atom potentials is relatively straightforward – the parameterization of such potentials is a much more significant problem.

5. CONCLUSIONS

There have been significant advances in our ability to quantify the forces between molecules from the distribution of charge within the molecule. Although we have not yet reached the stage where a definitive model potential can be routinely constructed for even small polyatomics, this goal looks achievable.

It would appear that, for many molecules, an accurate electrostatic model, plus relatively crude repulsion–dispersion terms, provides a sufficiently realistic model potential for simulating molecular crystal structures that the model potential is no longer a major barrier to progress in this field.

6. ACKNOWLEDGEMENTS

I would like to thank David S. Coombes for his assistance in producing this chapter, particularly the figures. The development of DMAREL was supported by the EPSRC under grants GR/J02872 and GR/F76718, with contributions from The Wellcome Foundation Limited. The computer resources were provided by EPSRC under GR/J31865.

7. REFERENCES

1. A. J. Kitaigorodsky, *Molecular Crystals and Molecules*, Academic Press, New York (1973); A. J. Pertsin and A. I. Kitaigorodsky, *The Atom–Atom Potential Method*, Springer Series in Chemical Physics 43, Springer-Verlag, Berlin (1987).
2. G. R. Desiraju, *Crystal Engineering. The Design of Organic Solids*, Elsevier, Amsterdam (1989).
3. G. Filippini and A. Gavezzotti, *Acta Cryst.*, **B49**, 868 (1993); A. Gavezzotti and G. Filippini, *J. Phys. Chem.*, **98**, 4831 (1994).
4. J. T. Brobjer and J. N. Murrell, *Molec. Phys.*, **50**, 885 (1983).
5. H.-J. Böhm and R. Ahlrichs, *Molec. Phys.*, **55**, 1159 (1985).
6. S. L. Price, *Molec. Phys.*, **58**, 651 (1986).
7. G. C. Maitland, M. Rigby, E. B. Smith and W. A. Wakeham, *Intermolecular Forces*, Clarendon, Oxford (1981).
8. J. A. Barker, R. A. Fisher and R. O. Watts, *Molec. Phys.*, **21**, 657 (1971).
9. I. C. Hayes and A. J. Stone, *Molec. Phys.*, **53**, 83 (1984); I. C. Hayes, G. J. B. Hurst and A. J. Stone, *Molec. Phys.*, **53**, 107 (1984).
10. A. J. Stone, in *Theoretical Models of Chemical Bonding*, Vol. 4, p. 103, edited by Z. B. Maksić, Springer, Berlin (1991).
11. S. L. Price, *Molec. Phys.*, **62**, 45 (1987).
12. F. Mulder and C. Huiszoon, *Molec. Phys.*, **34**, 1215 (1977).
13. U. C. Singh and P. A. Kollman, *J. Comput. Chem.*, **5**, 129 (1984).
14. K. B. Wiberg and P. R. Rablen, *J. Comput. Chem.*, **14**, 1504 (1993).
15. S. L. Price, *Molec. Simulation*, **1**, 135 (1988).
16. W. A. Sokalski and A. Sawaryn, *J. Chem. Phys.*, **87**, 526 (1987); J. P. Ritchie, *J. Amer. Chem. Soc.*, **107**, 1829 (1985); A. Pullman and D. Perahia, *Theor. Chim. Acta*, **48**, 29 (1978); F. Vigné-Maeder and P. Claverie, *J. Chem. Phys.*, **88**, 4934 (1988).
17. K. E. Laidig, *J. Phys. Chem.*, **97**, 12760 (1993).
18. Z. Berkovitch-Yellin, and L. Leiserowitz, *J. Amer. Chem. Soc.*, **102**, 7677 (1980).
19. A. J. Stone and M. Alderton, *Molec. Phys.*, **56**, 1047 (1985).
20. M. A. Spackman, *J. Chem. Phys.*, **85**, 6587 (1986).
21. S. L. Price, A. J. Stone and M. Alderton, *Molec. Phys.*, **52**, 987 (1984).

22. M. D. Ryan in *Modeling the Hydrogen Bond, ACS Symposium Series*, **569**, 36 (1994).
23. S. R. Cox and D. E. Williams, *J. Comput. Chem.*, **2**, 304 (1981).
24. S. L. Price, J. S. Andrews, C. W. Murray and R. D. Amos, *J. Amer. Chem. Soc.*, **114**, 8268 (1992).
25. A. J. Stone, *Molec. Phys.*, **56**, 1065 (1985).
26. C. R. LeSueur and A. J. Stone, *Molec. Phys.*, **83**, 292 (1994).
27. M. A. Spackman, *Chemical Reviews*, **92**, 1769 (1992).
28. B. M. Axilrod and E. Teller, *J. Chem. Phys.*, **11**, 299 (1943).
29. A. J. Stone and C. S. Tong, *Chem. Phys.*, **137**, 121 (1989).
30. S. L. Price and A. J. Stone, *Molec. Phys.*, **51**, 569 (1984).
31. S. L. Price, *Acta Cryst.*, **B42**, 388 (1986).
32. R. J. Wheatley and S. L. Price, *Molec. Phys.*, **69**, 507 (1990).
33. A. J. Stone, *Chem. Phys. Letts.*, **211**, 101 (1993).
34. K. Morokuma, *J. Chem. Phys.*, **55**, 1236 (1971).
35. S. C. Nyburg and C. H. Faerman, *Acta Cryst.*, **B41**, 274 (1985).
36. L.-Y. Hsu and D. E. Williams, *Inorg. Chem.*, **18**, 79 (1979); ibid., **18**, 2200 (1979).
37. S. L. Price and A. J. Stone, *Molec. Phys.*, **47**, 1457 (1982).
38. R. J. Wheatley and S. L. Price, *Molec. Phys.*, **71**, 1381 (1990).
39. E. D. Stevens, *Molec. Phys.*, **37**, 27 (1979).
40. G. R. Desiraju and R. Parthasarathy, *J. Amer. Chem. Soc.*, **111**, 8725 (1989).
41. S. L. Price, A. J. Stone, J. Lucas, R. S. Rowland and A. E. Thornley, *J. Amer. Chem. Soc.*, **116**, 4910 (1994).
42. R. J. Wheatley and J. B. O. Mitchell, *J. Comput. Chem.*, **15**, 1187 (1994).
43. R. J. Wheatley and W. J. Meath, *Molec. Phys.*, **80**, 25 (1993).
44. M. S. H. Ling and M. Rigby, *Molec. Phys.*, **51**, 855 (1984).
45. C. Millot and A. J. Stone, *Molec. Phys.*, **77**, 439 (1992).
46. P.-O. Åstrand, A. Wallqvist and G. Karlström, *J. Chem. Phys.*, **100**, 1262 (1994).
47. N. Gresh, *J. Comput. Chem.*, **16**, 856 (1995).
48. A. D. Buckingham and P. W. Fowler, *Canad. J. Chem.*, **63**, 2018 (1985).
49. A. C. Legon and D. J. Millen, *Acc. Chem. Res.*, **20**, 39 (1987).
50. G. J. B. Hurst, P. W. Fowler, A. J. Stone and A. D. Buckingham, *Int. J. Quant. Chem.*, **29**, 1223 (1986).
51. D. J. Wales, P. L. A. Popelier and A. J. Stone, *J. Chem. Phys.*, **102**, 5551 (1995).
52. S. L. Price and A. J. Stone, *J. Chem. Phys.*, **86**, 2859 (1987).
53. S. L. Price, F. Lo Celso, J. A. Treichel, J. M. Goodfellow and Y. Umrania, *J. Chem. Soc. Farad. Trans.*, **89**, 3407 (1993).
54. J. B. O. Mitchell, C. L. Nandi, J. M. Thornton, S. L. Price, J. Singh and M. Snarey, *J. Chem. Soc. Farad. Trans.*, **89**, 2619 (1993).
55. J. B. O. Mitchell and S. L. Price, *J. Comput. Chem.*, **10**, 1217 (1990).
56. C. H. Faerman and S. L. Price, *J. Amer. Chem. Soc.*, **112**, 4915 (1990).
57. P. D. Walker and P. G. Mezey, *J. Amer. Chem. Soc.*, **115**, 12423 (1993).
58. S. L. Price and A. J. Stone, *J. Chem. Soc. Farad. Trans.*, **88**, 1755 (1992).
59. C. A. Reynolds, J. W. Essex and W. G. Richards, *Chem. Phys. Letts.*, **199**, 257 (1992); T. Stouch and D. E. Williams, *J. Comput. Chem.*, **14**, 858 (1993).
60. S. L. Price, C. H. Faerman and C. W. Murray, *J. Comput. Chem.*, **12**, 1187 (1991).
61. U. Koch, P. L. A. Popelier and A. J. Stone, *Chem. Phys. Letts.*, **238**, 253 (1995); U. Koch and A. J. Stone, *J. Chem. Soc. Farad. Trans.*, **92**, 1701 (1996).
62. J. Perlstein, *J. Amer. Chem. Soc.*, **116**, 11420 (1994).
63. D. E. Williams and R. R. Weller, *J. Am. Chem. Soc.*, **105**, 4143 (1983).
64. D. J. Willock, S. L. Price, M. Leslie and C. R. A. Catlow, *J. Comput. Chem.*, **16**, 628 (1995).

65. P. L. A. Popelier and A. J. Stone, *Molec. Phys.*, **82**, 411 (1994).
66. F. H. Allen and O. Kennard, *Chem. Des. Automation News*, **8**, 31 (1993).
67. J. P. Ritchie, E. M. Kober and A. S. Copenhaver, *Proceedings of the 10th International Symposium on Detonation* (1993).
68. D. S. Coombes, S. L. Price, D. J. Willock and M. Leslie, *J. Phys. Chem.*, **100**, 7352 (1996).
69. F. H. Allen, O. Kennard, D. G. Watson, L. Brammer, A. G. Orpen and R. Taylor, *J. Chem. Soc., Perkin Trans. II*, S1 (1987).
70. P. C. Hariharan and J. A. Pople, *Theor. Chim. Acta*, **28**, 213 (1973).
71. CADPAC5: *The Cambridge Analytic Derivatives Package Issue 5*, A suite of quantum chemistry programs developed by R. D. Amos with contributions from I. L. Alberts, J. S. Andrews, S. M. Colwell, N. C. Handy, D. Jayatilaka, P. J. Knowles, R. Kobayashi, N. Koga, K. E. Laidig, P. E. Maslen, C. W. Murray, J. E. Rice, J. Sanz, E. D. Simandiras, A. J. Stone and M.-D. Su, Cambridge University (1992).
72. D. E. Williams and S. R. Cox, *Acta Cryst.*, **B40**, 404 (1984).
73. S. R. Cox, L.-Y. Hsu and D. E. Williams, *Acta Cryst.*, **A37**, 293 (1981).
74. A. J. Stone and C. S. Tong, *J. Comput. Chem.*, **15**, 1377 (1994).
75. S. L. Price in *Computer Modelling in Inorganic Crystallography*, edited by C. R. A. Catlow, Academic Press, London, in press.
76. J. S. Chickos, in *Molecular Structure and Energetics*, Vol. 2, edited by J. F. Liebman and A. Greenberg, VCH, New York, 1987, p. 67; and personal communication of updated tables.
77. M. G. Munowitz, G. L. Wheeler, S. D. Colson, *Molec. Phys.*, **34**, 1727 (1977).
78. D. E. Williams and T. L. Starr, *Comput. Chem.*, **1**, 173 (1977).

Chapter 3

Energetic Aspects of Crystal Packing: Experiment and Computer Simulations

ANGELO GAVEZZOTTI
Dipartimento di Chimica Strutturale e Stereochimica Inorganica, Università di Milano

AND

GIUSEPPE FILIPPINI
Centro CNR per lo Studio delle Relazioni tra Struttura e Reattività Chimica, Università di Milano

1. COMPUTER MODELING OF ORGANIC SOLIDS

This chapter is devoted to a critical analysis of the computational methods for the study of the energetics of organic crystals, together with their experimental counterparts. The basics of crystal thermodynamics and kinetics, and the techniques for the calculation of the energies of crystals and of solid-state processes will be described. The dynamical approach to computer simulations of organic molecules and solids, by molecular dynamics and Monte Carlo techniques, is dealt with in Chapters 4 and 6 of this book, and the simulation of phase transitions in small clusters is described in Chapter 5. Experimental and computational aspects of crystal growth and morphology are treated in Chapter 7.

To aid readability, no literature quotations are given in the text. Section 5 contains an annotated list of relevant papers and reviews.

Theoretical Aspects and Computer Modeling Edited by A. Gavezzotti

1.1 Solid-State Properties and Modeling Techniques

This section will deal with some mathematical aspects of the modeling of organic crystalline solids for application to computer simulations of chemical and physical properties of organic crystals. The basic task of these mathematical procedures is to produce some computer-oriented representation of a given object, repeated in space under the constraints of space-group symmetry.

In chemically oriented applications, the fundamental object is a molecule. Quite naturally, its best computer representation is a list of cartesian coordinates for the position of the nucleus of each atom in the molecule, in a given reference frame, much in the same style as the output of an ordinary single-crystal X-ray diffraction experiment. In fact, most computer simulations start with a molecular model based on that from X-ray work.

X-rays do not see nuclei, but electron clouds, so that X-ray 'nuclear' positions are in fact the centroids of the electron distribution around each atom. This is satisfactory for all atoms except hydrogens, whose positions are determined with much lower accuracy than those of heavier atoms, so that unreasonable geometries often occur. Neutron diffraction studies locate hydrogen nuclei with accuracy comparable to that of the positions of heavier nuclei; unfortunately, neutron diffraction is still little used by crystallographers. Hydrogens are abundant in organic molecules, and have a very specific and important role in the energetics of crystal packing, so that their consistent location is relevant for simulations. Hydrogen coordinates from X-ray studies cannot be used in solid-state calculations, and hydrogen atom positions are usually adjusted and sometimes located by geometrical rules from the positions of the surrounding heavier atoms. Crystallographic studies use a cell reference system that in general is not orthogonal and has different unit vectors along each direction. This adds a small supplementary technical problem in computer simulations, but a few simple matrix manipulations can take care of the problem.

Once a full set of atomic coordinates for the asymmetric unit has been obtained ($\mathbf{x}^0$), the crystal can be represented by a list of nuclear coordinates, in which the asymmetric unit has been repeated in space according to space-group symmetry elements. The matrix–vector representation of space groups is perfectly adequate for this task; any symmetry-related asymmetric unit can be represented by a set of coordinates, $\mathbf{x}_j$, where

$$\mathbf{x}_j = \mathbf{M}_j\mathbf{x}^0 + \mathbf{t}_j, \qquad (1)$$

$\mathbf{M}_j$ and $\mathbf{t}_j$ being a matrix–vector pair. Integers can be added to the $\mathbf{t}_j$ components, so that the crystal can extend to infinity, or at least to a cluster of as many molecules as desired.

Once the positions of the nuclei of all atoms in the crystal cluster are known, the crystal packing is completely determined. Intermolecular bond distances and angles can be calculated, and used in statistical studies of crystal-packing geometry using databases. The lattice energy can be calculated using parametric force fields which depend only on nuclear geometry, or by quantum-chemical methods. Total energies and their first and second derivatives with respect to a chosen set of crystal coordinates can be calculated. The analytical calculation of these derivatives is somewhat complicated, but they can also be approximated by finite differences.

In some quantum-chemical applications, the object to be repeated by space-group symmetry is an atomic orbital. The symmetry-adapted linear combinations of atomic orbitals under translational symmetry are called 'Bloch functions'. The simplest form of such functions, for a monodimensional lattice, is

$$\Psi = \sum_n \exp(\mathrm{i}\ \mathbf{k}\ n\ a)\chi_n \tag{2}$$

where n is the index that labels lattice points (to infinity), a is the lattice spacing, χ_n is an atomic orbital at the n-th lattice site, and $\mathbf{k}$ is a wave vector, whose unique values are in the range $0 < \mathbf{k} < \pi/a$. Among other things, $\mathbf{k}$ counts the number of nodes in the function. An infinite set of linear combinations Ψ are generated in order of decreasing stability with increasing number of nodes, and this construction is called an 'energy band'. Intuitively, a parallel can be drawn with the p-orbitals of a polyene, whose number of conjugated double bonds tends to infinity; the linear combinations increase in number to infinity, and the spacing between them decreases to an infinitesimal amount.

The ordinary integrals from the Hartree–Fock treatment are then calculated over periodic, rather than simple atomic orbitals. Of course, the presence of different atomic orbitals over various atoms and of space-group symmetry within the cell make the mathematical treatment rather complicated, but the basic principle does not change. The input to such calculations consists of the nuclear coordinates of atoms in the asymmetric unit, and of the atomic orbital basis set. As for the LCAO-MO method in quantum chemistry for isolated molecules, the crystal wavefunction and total energy can be calculated, so that this periodic-orbital approach can be used to study the electronic properties of a crystal. For an isolated molecule, the LCAO-MO treatment leads to a molecular orbital energy spectrum; in the crystalline, periodic case, its equivalent is a density of electronic states (the number of states between E and $\mathrm{d}E$) taken over the energy band.

In the calculation of lattice vibrations, the object to be periodically reproduced in space is an atomic displacement from equilibrium or a rigid-body libration near equilibrium. The operator that performs this periodicization is conceptually similar to the one used in Bloch functions, and a typical periodic vibrational coordinate, for rigid-body molecules, has the form

$$A = A^0 \exp(\mathrm{i}\ \mathbf{k}\ \mathbf{R}) \exp(-\mathrm{i}\ \omega\ t) \tag{3}$$

where A^0 is the vibration amplitude, and $\mathbf{R}$ is a vector joining molecular centers of mass. In analogy to Bloch functions, the first exponential is the periodic operator, and $A^0 \exp(-i\ \omega\ t)$ is the elementary vibration to be made periodic.

An infinite number of vibrations with an increasing number of nodes and energy are obtained as linear combinations of the vibrational coordinates. The Newton equations of motion are solved using the periodic displacements described above; the harmonic approximation consists of expanding the potential energy in a series of powers, neglecting all terms above second-order. The dynamical matrix, containing the force constants, is calculated from the second derivatives of the potential energy.

Aside from the mathematical complexity of the treatment, the whole procedure is conceptually similar to the GF-matrix method for molecular vibrations: both lead to the solution of a vibrational eigenvalue problem, and to the determination of normal coordinates – in the case of crystals, the lattice-vibrational modes. If the displacement coordinates refer to each single atom separately, the whole intra- and intermolecular force field must be used; rigid molecules are defined as having intramolecular vibrations that do not couple with intermolecular ones, and in that case vibrational displacement coordinates only refer to motion of the center of mass and rotations of the whole molecular object (six degrees of freedom per molecule). This approximation greatly reduces the complexity of the calculations.

In isolated molecules, the procedure leads to a discrete number of vibrational frequencies, from which the vibrational part of the thermodynamic functions (enthalpy, entropy, specific heat) can be calculated by statistical thermodynamics. In crystalline, infinitely periodic systems, the equivalent of the vibrational spectrum is a vibrational band, over which a density of vibrational states can be calculated. This density of states can be used to calculate the vibrational part of the crystal thermodynamic functions. If the molecule is rigid, internal contributions are not considered and only external (intermolecular) contributions are obtained.

The first derivatives of the crystal potential are crystal forces. They must be zero for a crystal structure in mechanical equilibrium, and act as restoring forces if the crystal structure is mechanically perturbed. They are of course highly anisotropic, and their calculation opens the way, in principle, to the calculation of mechanical properties of solids, such as stress, strain or compressibility moduli. The computational procedures for these calculations require an evaluation of these forces as a function of displacements from equilibrium positions, and, from these, of the corresponding tensorial quantities.

The knowledge of the crystal-packing geometry and of the molecular wavefunction and electronic polarizability allows, in principle, the calculation of optical properties of crystals, and, if the wavefunction is spin-sensitive, also of magnetic properties. The response of the crystal electron density and charge distribution to the field of an incoming electromagnetic radiation may be non-linear, and in that case frequency variations occur. This property, even for organic crystals, is important in non-linear signal processing. A specialized literature, mostly in solid-state physics, deals with these issues. The calculations are, however, more easily, and hence more frequently, performed for the much more polarizable inorganic substances.

Organic crystals, like all crystals, contain an unknown amount of disorder and defects, in the form of vacancies (missing molecules, with formation of voids), impurities (unwanted guest molecules substituting for molecules of the original compound), rotational and translational diffusion (molecules or groups of molecules undergoing reorientations or translation within the crystal), dislocations and domains (the crystal being formed of a mosaic of sizable subunits within which order is maintained). Molecular reorientations are fairly common in organic crystals, and very frequent near the melting point. The ease of reorientation strictly depends on molecular shape; for example, even very large, but flat and round, condensed aromatic hydrocarbon molecules reorient in their crystal at low temperature. The time scale of these motions is such that they can be detected by NMR.

The computer modeling of each of these disorder phenomena, let alone their combinations, is in general rather difficult. It is relatively easy to calculate the energy barrier against reorientation of a single molecule in a rigid crystal environment; problems arise when the modeling must include relaxation of the environment, which, understandably, is a most likely occurrence.

A more sparse, but not entirely uncommon occurrence is solid-state reactivity, or a true chemical reaction taking place within the crystal structure (this is much more common in inorganic crystals). The modeling of this phenomenon is not simple, since in principle the influence of the environment, and its relaxation or transformations, must be taken into account even when studying reaction at the level of one single reacting molecule. Experimental evidence shows that solid-state reaction seldom, if ever, occurs by a single-crystal to single-crystal mechanism. Growth of product crystal seeds within the reactant structure usually leads to its destruction, and to the formation of, at best, polycrystalline products. A detailed computational description of the whole process seems at present well beyond the reach of modeling techniques.

1.2 Crystal Potentials

1.2.1 Empirical Force Fields

Early derivations of empirical intermolecular force fields for organic crystals were based on a consideration of the quantum-mechanical origin of the dispersion–repulsion forces, and resulted in analytical formulations including, essentially, exponential and inverse-power functions of interatomic intermolecular distances, total energies being taken as lattice summations over such terms. Subsequent revisions and extensions tended to take these analyses for granted, proceeding to empirical fits of the necessary parameters using newly determined experimental properties, leaving the analytical form of the functions unchanged, and with little or no rethinking of the original principles. This method led, however, to a considerable success of what was eventually called 'the atom–atom approach', whose strong points are simplicity and easy application to fast computer calculations.

Typical complete (intra- and intermolecular) force fields for large-scale computer applications to static and dynamical calculations of the energies of condensed phases include bond stretching, bond angle bending, and several types of torsional contributions, as in force fields for gas-phase molecules, plus intermolecular terms of the 6-exp or 6-12 type. Coulomb-type contributions are included using supplementary terms in the inverse first power of interatomic (intermolecular and, in most cases, also intramolecular) distances. Charge parameters have to be supplied in large numbers, for each atom type; these are mostly obtained from quantum-chemical calculations, by several types of decomposition of the molecular charge density, ranging from simple Mulliken population analysis to the fitting of the electrostatic potential by either point charges or distributed dipoles. More on this last subject is to be found in Chapter 2 of this book.

The philosophy and practical aspects of fitting techniques for the derivation of molecular force fields have been reviewed extensively (see also Chapter 4). Recent developments have sometimes been more towards better graphics or better energy-surface search methods, rather than towards an improvement of the force fields themselves, and some commercial molecular modeling packages make great use of 'default' parameters of dubious origin. An appreciation of the relationships between the derivation of the force field and the chemical problem to which the field is to be applied is sometimes sidestepped, and there is danger of widespread uncritical application.

Since the focus of this book is on crystals, an example will be given of an intermolecular force field explicitly calibrated upon, and dedicated to, the organic solid state. Its essential features can be summarized as follows:

(a) the functional form is as simple as possible, namely, the Buckingham-type atom–atom formulation: $E(R_{ij}) = A\cdot\exp(-BR_{ij}) - CR_{ij}^{-6}$, where R_{ij} is the distance between any two atoms in different molecules. No R^{-1} terms are used in a first approximation; this allows an easy application to large-scale computer simulations;
(b) the number of independent A, B, C parameters is very small, but cross-interaction parameters (the key ones in organic crystal-packing) are separately optimized, instead of using standard averaging techniques; this procedure recovers a large part of the coulombic interaction in the deeper-than-usual potential wells between atoms of different electronegativity, and allows a coherent treatment of hydrogen bonding;
(c) supplementary coulombic terms can be introduced, and even rescaled point

charges from Mulliken population analysis over an extended Hückel wave-function have proven adequate.

The fitting procedure uses a very large number of experimental data, that is, hundreds of known crystal structures and heats of sublimation for all the most common molecular types and functionalities of organic chemistry. The position of the minimum in the potential curve is determined using a statistical analysis of observed intermolecular contacts in selected samples of crystal structures retrieved from the Cambridge Structural Database, while the well depths are scaled by a comparison between experimental heats of sublimation and calculated packing energies.

The potentials so determined are collected in Table 3.1. For application to

Table 3.1 Atom–atom potential parameters: $E = A \cdot \exp(-BR_{ij}) - CR_{ij}^{-6}$

Interaction	A^a	B^b	C^c	ε^d	R^e
H...H	5774	4.01	26.1	0.010	3.36
H...C	28 870	4.10	113	0.049	3.29
H...N	54 560	4.52	120	0.094	2.99
H...O	70 610	4.82	105	0.121	2.80
H...S	64 190	4.03	279	0.110	3.35
H...Cl	70 020	4.09	279	0.120	3.30
C...C	54 050	3.47	578	0.093	3.89
C...N	117 470	3.86	667	0.201	3.50
C...O	93 950	3.74	641	0.161	3.61
C...S	126 460	3.41	1504	0.217	3.96
C...Cl	93 370	3.52	923	0.160	3.83
N...N	87 300	3.65	691	0.150	3.70
N...O	64 190	3.86	364	0.110	3.50
O...O	46 680	3.74	319	0.080	3.61
O...S	110 160	3.63	906	0.189	3.72
O...Cl	80 855	3.63	665	0.139	3.72
S...S	259 960	3.52	2571	0.445	3.83
Cl...Cl	140 050	3.52	1385	0.240	3.83
HB...O (amides)	3 607 810	7.78	238	4.0	1.80
HB...O (acids)	6 313 670	8.75	205	7.0	1.60
HB...O (alcohols)	4 509 750	7.78	298	5.0	1.80
HB...N (=N—H...N)	7 215 600	7.78	476	8.0	1.80
HB...N (—NH_2...N)	1 803 920	7.37	165	2.0	1.90
HB...O (NH_2—NO_2)	1 315 000	7.78	86.7	1.46	1.80
F...F	40 850	4.22	135	0.07	3.20
C...F	46 990	3.84	279	0.08	3.50
H...F	15 358	4.11	59	0.03	3.30
Si...Si	232 473	3.21	3953	0.40	4.21
Si...C	112 094	3.34	1512	0.19	4.04
Si...H	36 637	3.61	321	0.07	3.72

[a]kcal/mol. [b]$Å^{-1}$. [c]kcal/(mol·$Å^{-6}$). [d]Potential well depth (kcal/mol). [e]Distance at the minimum (Å). $\lambda = 13.5$ (14.0 for hydrogen-bonding functions).

general organic crystals, with many functional groups and complex interactions for which potential parameters were not explicitly optimized, a number of ad hoc adjustments and averages must be postulated. Table 3.2 shows the calculated lattice energies for polyfunctional, complex organic molecules not included in the calibration set, in comparison with experimental sublimation heats. The agreement is encouraging, especially considering the wide chemical range of application.

Table 3.2 Packing energies calculated with the potentials in Table 3.1, and experimental heats of sublimation for a selected sample of organic crystals, not included in the calibration (kcal/mole units).

CSD ref. code	Substance	PE	ΔH_s	% $\lvert\Delta H_s - \text{PE}\rvert/\Delta H_s$
AMBACO05	*o*-Aminobenzoic acid (orthorhombic form)	27.9	26.7	4.5
AMBACO06	*o*-Aminobenzoic acid (monoclinic form)	26.8	26.7	0.4
AMBNAC01	*p*-Aminobenzoic acid	28.1	27.7	5.8
AZTHYM10	6-Azathymine	28.2	26.9	4.8
BALNIN01	β-Alanine	25.3	25.1	0.8
BIFPEL	*t*-Butyl carbamate	19.7	22.5	12.4
BIPYRL	2,2′-Bipyridyl	22.1	19.6	7.6
BZTRAZ	Benzotriazole	27.1	23.7	14.3
CYTSIN01	Cytosine	32.5	35.2	7.7
DACETA	trans-cis-Diacetamide	18.3	17.5	4.6
DLMETA03	DL-methionine β form	28.1	32.0	12.2
DTURAC	2,4-Dithiouracil	29.6	28.6	3.5
ECARBM10	Ethyl carbamate	19.4	18.6	4.3
GEBTUC	Hypoxanthine	34.9	37.8	7.7
HXQUIN03	8-Hydroxyquinoline	27.8	21.4	29.9
HXQUIO	2-Hydroxyquinoline	25.4	28.5	10.9
MALOAM	Malonamide	30.1	30.2	0.3
MBNZAM10	*N*-Methylbenzamide	22.4	17.9	25.1
METURA03	*N*-Methyluracil	23.2	26.9	13.8
OCHTET03	1,3,5,7-Tetranitro-1,3,5,7-tetra-azacyclo-octane (δ form)	29.9	38.7	22.7
OCHTET12	1,3,5,7-Tetranitro-1,3,5,7-tetra-azacyclo-octane (β form)	34.9	41.9	16.7
PICRAC	Picric acid	28.3	25.1	12.7
PROLIN	L-Proline	24.0	30.4	21.1
QQQBRD02	Hexanitroethane	20.3	16.9	20.1
THACEM01	Thioacetamide at 15 K	20.4	19.9	2.5
THCYTO10	Thiocytosine	35.7	37.8	5.6
TURCIL01	2-Thiouracil	28.5	30.9	7.8
URACIL	Uracil	26.3	31.3	16.0

1.2.2 Quantum-Chemical Calculations

A straightforward way of calculating crystal energies and relative stabilities for organic crystal structures is to build a many-molecule cluster, and to perform a quantum-chemical calculation on this supramolecular aggregate. The limitations of this approach, in both theory and practice, are evident: very sophisticated quantum-chemical calculations are required for the reproduction of weak intermolecular interactions, and this severely restricts the size of the molecule and of the cluster, as far as computing times are concerned. Cluster-truncation effects reduce the significance of those calculations for real crystals.

The use of periodic orbitals reduces to some extent these restrictions. Since, however, the extension of the calculation schemes beyond the Hartree–Fock level is also computationally demanding, the weak correlation interactions proper of organic crystals are once again beyond reach. A much more productive application of these methods is to inorganic or metallic crystals, in a much stronger intermolecular or interatomic regime. Density-functional methods apparently reduce the costs of the introduction of correlation energy in the calculations, and some applications to crystals have been proposed.

Presently, the applicability of quantum-chemical calculations increases as the molecular recognition problem becomes more localized. Stronger local interactions, like hydrogen bonding, are better reproduced, while the packing of hydrogen-bonded oligomers or chains, dictated by weaker interactions, is more elusive. On the other hand, pairwise-additive empirical schemes are not able to reproduce cooperative or delocalization effects, such as the non-additivity of the energy of multiple hydrogen bonds. In this respect, a promising strategy is the use of high-quality quantum-chemical calculations on small systems to sample the basic crystal interactions, and the exploitation of the energetic results for the calibration of empirical force fields, or of the wavefunctions for an accurate evaluation of the electrostatic crystal potential. The reader is referred to Chapter 2 in this book for more on this last subject.

2. THERMODYNAMICS

It is often stated that modern organic chemistry is a molecular science, and that an understanding of the structure and energetics of molecules is the key to understanding the bulk chemical properties of a compound. This point of view certainly carries much promise for future developments, but should not be taken as a suggestion that classical thermodynamics and kinetics are doomed to oblivion. Any molecular prediction has to be eventually checked against bulk temperature- or time-dependent measurements, which bring back those two glorious and indispensable disciplines to the forefront of chemical understanding.

The basic thermochemical measurement on solids (as well as, for that matter, on other phases) is the heat capacity at constant pressure, C_P. Enthalpy and entropy are then accessed using the basic relationships:

$$\Delta H(T_2 - T_1) = \int C_P(T)\,dT \tag{4}$$

$$\Delta S(T_2 - T_1) = \int [C_P(T)/T]\,dT$$

In the following, the subscripts or superscripts c, l and v refer to solid (crystalline), liquid and vapor, respectively, and ΔX_i^j is the variation in the thermodynamic function X on going from state i to state j.

Other important calorimetric determinations include heats of combustion, giving access to the enthalpy of formation of the solid, $\Delta H_f(c)$; heats of melting, ΔH_c^l, which can be measured directly by calorimetry from the areas of the relevant peaks; and heats of sublimation, ΔH_c^v, which are measured by several different techniques, but most often by Clausius–Clapeyron treatment of vapor pressure measurements:

$$\ln (P_2/P_1) = [\Delta H_c^v/R](1/T_1 - 1/T_2) \tag{5}$$

The vapor pressures of solids being very low, this last measurement is critical, and also rather tedious.

The following sections illustrate some aspects of the computer simulation of these key quantities for organic crystals.

2.1 The Solid–Vapor Equilibrium

The sublimation enthalpy, ΔH_c^v, can be assumed equal to half of the calculated packing potential energy (PPE), or packing energy (PE):

$$\Delta H_c^v = \tfrac{1}{2}\text{PPE} = \tfrac{1}{2}\sum E(R_{ij}) = \text{PE} \tag{6}$$

This expression is not quite correct, since the calculation refers to an undetermined temperature, while sublimation enthalpies are sensitive to temperature; besides, the calculation yields in principle the sublimation internal energy, ΔU_c^v, that is, the energy difference between a molecule in the crystal and a molecule in the vapor, both at rest. The equipartition kinetic energy for the vapor is $3RT$ (from translation and rotation), and the vibrational and rotational equipartition energy for the crystal with both kinetic and potential energy contributions is $2 \times 6 \times \frac{1}{2}RT = 6RT$; contributions from intramolecular vibrations are assumed to be the same in the solid and in the vapor, and are neglected. Therefore:

(a) neglecting molecular motions:

$$\Delta H_c^v = \Delta U_c^v + \Delta(PV) = \Delta U_c^v + P\Delta V = \Delta U_c^v + P{\cdot}V(\mathrm{v}) = \Delta U_c^v + RT \quad (7)$$

(b) including molecular motions:

$$\Delta H_c^v = \Delta U_c^v + RT + \Delta U_{\mathrm{vib}} = \Delta U_c^v + RT + (3RT - 6RT) = \Delta U_c^v - 2RT \quad (8)$$

Also Eq. (8) is not quite correct, since several other factors need be considered (see Appendix, Section 7 of this chapter). These issues can be sidestepped if the force field is calibrated to directly reproduce ΔH_c^v at 300 K; in that case, comparisons between experiment and calculation refer to Eq. (6). When R^{-1} terms are not included, lack of convergence in the lattice sums of Eq. (6) is compensated by using a 10 Å cutoff radius and multiplying the calculated PE by 1.05; with these terms, reciprocal lattice sums are used to force convergence. When the molecule is conformationally flexible, the intramolecular energy difference between the solid and the vapor must be added. This is a critical point, especially when intra- and intermolecular force fields have been derived separately and are therefore not consistent.

The calculation of PPE is straightforward when the crystal structure is known from single-crystal X-ray diffraction, and takes a few CPU seconds on any modern workstation. Within the above-discussed limitations, it can be stated that the ΔH_c^v of an organic compound whose crystal structure is known at room temperature can be calculated in an easy and inexpensive way with a high degree of reliability. When the crystal structure is not known, the heat of sublimation can also be estimated, using an approximation described in Section 4 of this chapter.

The sublimation entropy can be calculated as follows (at standard pressure, usually 1 atm):

$$(\Delta S_c^v)^0 = S^0(\mathrm{int, v}) + S^0(\mathrm{rot, v}) + S^0(\mathrm{trasl, v}) - S^0(\mathrm{int, c}) - S^0(\mathrm{ext, c}) \quad (9)$$

where the suffixes 'int' and 'ext' refer to the intra- and intermolecular vibrational contributions, respectively (vibrational coupling of internal and external modes is neglected). The terms for the vapor can be evaluated by standard formulas:

$$S^0(\mathrm{trasl, v}) = \tfrac{3}{2}R\ln(M) + \tfrac{5}{2}R\ln T - 9.67 \quad \mathrm{J/(K.mole)} \quad (10)$$

$$S^0(\mathrm{rot, v}) = \tfrac{3}{2}R + R\ln[(1/\sigma)0.015\,02T^{3/2}(I_xI_yI_z)^{1/2}] \quad \mathrm{J/(K.mole)} \quad (11)$$

where M is the molecular weight in grams, σ is the symmetry number, and I_x, I_y, I_z are the moments of inertia in amu.Å^2. Note that $S^0(\mathrm{ext, c})$ can be evaluated from lattice-dynamical calculations as described in Section 1.1 in this chapter. ΔS_c^0 can then be readily calculated, assuming $S^0(\text{int, v}) = S^0(\text{int, c})$, if no large conformational rearrangements occur between the two phases.

In the same way, the ΔC_P between vapor and solid can be calculated (at 1 atm standard pressure) as:

$$\Delta C_P = [3R + C_P(\text{int, v})] - [C_P(\text{int, c}) + C_P(\text{ext, c})] \tag{12}$$

where the first term arises from equipartition in the vapor, and, as above, the internal vibrational contributions in the solid and in the vapor are assumed to be equal. $C_P(\text{ext, c})$ is obtained, like the corresponding entropy term, from lattice-dynamical calculations. Then:

$$\Delta H_c^v(T) = \Delta H_c^v(298) + \int_{298}^{T} (\Delta C_P)\,dT \tag{13}$$

Once $(\Delta S_c^v)^0$ and ΔH_c^v are available as functions of temperature, $(\Delta G_c^v)^0$ can be obtained at any temperature, and hence the vapor pressure of the solid, P, can also be obtained as a function of temperature, by writing the equilibrium condition (where P^0 is the standard pressure, usually 1 atm):

Table 3.3 Sublimation data for organic compounds (energies in kcal/mole, entropies in cal/(k·mole), pressures in atm) at 300 K

Compound	ΔH_s		S^0(v)	S^0 (ext, s)	ΔS_s^0		Vapor pressure	
	exp	calc			calc	exp	calc	exp
Benzene	10.6	10.6	59.5	26.3	33.2		0.30	0.18[a]
Naphthalene	17.4	17.6	66.5	26.9	39.6	40.3	5×10^{-5}	1×10^{-4} [b]
Phenanthrene	22.1	22.6	71.0	30.1	41.0	43.3	2×10^{-8}	2×10^{-7} [b]
Anthracene	25.0	24.2	69.7	28.1	41.6	47.0	2×10^{-9}	7×10^{-9} [b]
Chrysene	31.3	30.3	73.4	28.3	45.0	53.7	4×10^{-13}	6×10^{-12} [b]
1,2-Benzanthracene	29.5	27.8	74.9	31.2	43.7	52.7	1×10^{-11}	7×10^{-11} [b]
Biphenyl	19.5	21.1	68.5	28.8	39.7		2×10^{-7}	1×10^{-5} [c]
9,10-Anthraquinone	27.1	29.8	70.9	27.7	43.3		4×10^{-13}	2×10^{-10} [a]
Benzoic acid	21.3	22.3	68.8	24.5	44.3		2×10^{-7}	1×10^{-6} [a]
Phthalic anhydride	21.2	21.2	69.1	27.3	41.8		4×10^{-7}	8×10^{-7} [a]
Coumarin	20.6	22.0	70.3	27.5	42.8		2×10^{-7}	—
Tetracyanomethane	14.6	14.2	68.9	26.3	42.6		8×10^{-2}	—
Phenazine	23.2	26.4	69.6	27.7	41.9		5×10^{-11}	5×10^{-8} [d]
p-Dichlorobenzene	15.5	16.0	67.3	30.3	37.0		2×10^{-4}	2×10^{-3} [a]
p-Nitrotoluene	18.9	20.6	69.9	30.6	39.2		3×10^{-7}	8×10^{-6} [e]
DDT	28.2	29.6	78.7	33.0	45.8		2×10^{-12}	7×10^{-10} [a]
Benzamide	24.3	25.2	68.8	24.0	44.8		2×10^{-9}	—

[a] A. Hamilton-Jones, *J. Chem. Eng. Data*, **5**, 196 (1960). [b] C. G. De Kruif, *J. Chem. Thermod.*, **12**, 243–248 (1980). [c] R. D. Chirico, S. E. Knipmeyer, A. Nguyen and W. V. Steele, *J. Chem. Thermod.*, **21**, 1307 (1989). [d] D. M. McEachern, O. Sandoval and J. C. Iniguez, *J. Chem. Thermod.*, **7**, 299 (1975). [e] C. Lenchitz and R. W. Velicky, *J. Chem. Eng. Data.*, **15**, 401 (1970).

$$\mu^*(\mathrm{c}) = \mu^0(\mathrm{v}) + RT\ln(P/P^0) \tag{14}$$

$$\mu^*(\mathrm{c}) - \mu^0(\mathrm{v}) = -(\Delta G_\mathrm{c}^\mathrm{v})^0 = RT\ln P \tag{15}$$

Table 3.3 shows some results of vapor pressure calculations for a selection of organic crystals. Order-of-magnitude agreement appears, and the largest deviations are for crystals whose calculated PE differs considerably from the measured sublimation enthalpy. Similarly to sublimation enthalpies and entropies, and within the same limitations, the vapor pressure of organic crystals whose structure has been determined by X-rays can also be calculated with reasonable approximation.

2.2 The Solid–Liquid Equilibrium

In solid–vapor (and liquid–vapor) equilibria the chemical potential of the vapor is expressed in terms of activity (the vapor pressure, which is in fact the 'equilibrium constant' of the process), while condensed phases are by definition at unit activity. Therefore, in the solid–liquid equilibrium, at constant pressure, the free energy of melting can be written as:

$$\Delta G_\mathrm{c}^\mathrm{l}(T) = \mu_\mathrm{l}(T) - \mu_\mathrm{c}(T) \tag{16}$$

Equilibrium is reached at the melting temperature, $\Delta G_\mathrm{c}^\mathrm{l}(T_\mathrm{m}) = 0$, and

$$T_\mathrm{m} = \Delta H_\mathrm{c}^\mathrm{l}/\Delta S_\mathrm{c}^\mathrm{l} \tag{17}$$

The a priori evaluation of the melting temperature by the above formula requires in principle a separate (and accurate) evaluation of the enthalpies and entropies of the solid and of the liquid as a function of temperature, which is at present an impossible task. In practice, the melting temperature and the enthalpy of melting can be readily measured by calorimetry, although the experiment is sometimes made difficult by mesomorphic phase transitions in proximity of the melting point. Assuming that the process is an equilibrium one (a condition that can be approached by careful instrumental settings) the melting entropy can be evaluated as

$$\Delta S_\mathrm{c}^\mathrm{l} = \Delta H_\mathrm{c}^\mathrm{l}/T_\mathrm{m} \tag{18}$$

The molecular mechanisms of melting are not known, at least for complex organic molecules. To illustrate some of the principles involved, a simplified computational model for the calculation of melting enthalpies from crystal data will be presented. Let the crystal structure be described, as usual, by one reference molecule and N surrounding molecules in known positions. To simulate the expansion that occurs on melting, typically 10–15% for organics, cell dimensions are increased isotropically, leaving cell angles unaltered. A zero-level computational model assumes that the melting enthalpy is just the difference in packing energy (PE) between the original and the expanded structure (this model ignores rotational and translational motions in the liquid, as well as their correlation):

$$\Delta H_c^l = \tfrac{1}{2}[\text{PEE} - \text{PPE(expanded lattice)}] \tag{19}$$

A first-level model assumes that the surrounding molecules can take any orientation with respect to the reference one, while their centers of mass remain fixed, or are slightly displaced away from the reference molecule, as a result of lattice expansion. Molecular rotations are performed by scanning the full range of three Euler angles. For the j-th surrounding molecule, let E_j^* be the interaction energy with the reference molecule in the crystal, and E_j^0 the optimum interaction energy over the rotations; for each rotation k, let E_{jk} be the interaction energy between the rotated surrounding molecule and the reference one. Then, the average energy of interaction between the reference molecule and the j-th surrounding one, $\bar{E}_j$, can be calculated as follows:

$$E_j = \sum_k E_{jk} \exp[-(E_{jk} - E_j^0)/RT] \tag{20}$$

$$Q_j = \sum_k \exp[-(E_{jk} - E_j^0)/RT] \tag{21}$$

$$\bar{E}_j = E_j/Q_j \tag{22}$$

T may be taken as 300 K as a reasonable guess but without much theoretical basis for the choice. The expectation value of the potential energy of the liquid is

$$\langle E_l \rangle = \sum_j \bar{E}_j \tag{23}$$

while, of course, $\langle E_c \rangle = \langle E_j^* \rangle = \sum_j E_j^* = \text{PPE}$. The difference between the average 'packing' energy of the liquid and the packing energy of the solid is then equated to the melting enthalpy at 300 K:

$$\Delta H_c^l = \tfrac{1}{2}(\langle E_l \rangle - \langle E_c \rangle) \tag{24}$$

The above model is conceptually similar to the so-called positional melting model; however, it ignores translational disordering and vacant lattice sites in the premelted crystal (on the concentration of which very little is known for molecular crystals). It also ignores intermolecular correlation between the molecular rotations, that is, it assumes essentially free molecular rotation in the liquid. Both zero- and first-level models neglect, among other things, differences in molecular conformation between the solid and the liquid, and hence can be applied only to rigid molecules.

Table 3.4 shows some numerical results. Both models, although very crude, yield melting enthalpies of the correct order of magnitude. The first-level model gives good results for hydrocarbons (for naphthalene, even the experimental

Table 3.4 Calculated and experimental heats of melting (kcal/mole) and entropies of melting (cal/K·mole)

Compound	ΔH_m, calc[a]		ΔH_m, exp	S^0(ext, l)
	+9%	+16%		
Benzene	2.2	2.8	2.35	11.1
	0.8	1.6		
Naphthalene	3.3	4.3	4.61	10.8
	1.2	2.4		
Anthracene	3.0	4.5	6.89	9.5
	1.8	3.6		
Phenanthrene	3.6	4.9	4.45	10.8
	1.8	3.5		
Succinic anhydride	2.0	2.8	4.87	
	0.4	1.3		
Anthraquinone	2.5	4.6	7.80	
	2.2	4.3		
Benzoic acid	0.4	1.1	4.10	
	—	—		
Benzoquinone	2.3	3.4	4.41	
	0.8	1.8		
Biphenyl	3.5	4.9	4.44	
	—	—		
Benzil	2.3	4.2	4.65	
	2.3	5.0		
n-Dodecane	1.5	3.0	8.70	
	—	—		
o-Aminobenzoic acid	3.3	6.1	4.86	
	6.3	9.8		
p-Chlorobenzoic acid	−0.8	0.8	7.70	
	1.7	3.1		
m-Chlorobenzoic acid	3.4	6.0	5.70	
	6.3	8.5		
m-Dinitrobenzene	3.6	5.2	4.15	
	2.3	4.2		
p-Dinitrobenzene	2.9	4.6	6.72	
	1.9	3.9		
p-Nitrobenzoic acid	1.9	3.8	8.82	
	6.0	13.9		

[a]Calculated value at shown % expansion of the lattice; for each compound, first row, Eq (24), second row, Eq. (19).

volume expansion is quantitatively reproduced), and a few unsatisfactory results for hydrogen-bonded crystals, where the neglect of correlation between molecular motions allows unrealistic attractive contacts between rotated molecules and the reference one, yielding negative heats of melting. When the lattice is suitably expanded, the no-correlation assumption becomes more and more realistic, and the calculated values are more reasonable. The zero-level model apparently owes its partial success to the cancellation of the effects (both neglected) of rotational/translational disordering and of correlation.

An approximate expression for the entropy of melting can be written as follows:

$$(\Delta S_{\mathrm{c}}^{\mathrm{l}})^0 = S^0(\mathrm{l}) - S^0(\mathrm{c}) = S^0(\mathrm{int,l}) + S^0(\mathrm{rot,l}) + S^0(\mathrm{ext,l})$$
$$-S^0(\mathrm{int,c}) - S^0(\mathrm{ext,c}) + \Delta S^0(\mathrm{pos}) \tag{25}$$

where $S^0(\mathrm{int,l}) - S^0(\mathrm{int,c}) = 0$, as for sublimation, and under the same conditions; $S^0(\mathrm{rot,l})$ can be taken as equal to the rotational entropy of the vapor if free rotation in the liquid is assumed, and can be calculated as described above; $S^0(\mathrm{ext,c})$ is also calculated as described above; and $\Delta S^0(\mathrm{pos})$ is the entropy difference due to positional disordering of the molecular centers of mass, which should not be too different from the monoatomic gas value, 3 cal·K^{-1}·mole^{-1}. Therefore, the only term in Eq. (25) that cannot be readily calculated (or measured) is $S^0(\mathrm{ext,l})$, or the entropy contribution from intermolecular 'vibrations' in the liquid. Recalculating this term from measured overall entropies of melting for rigid molecules gives the results shown in the last column of Table 3.4.

The thermodynamic functions for melting could in principle be calculated by molecular dynamics or Monte Carlo simulations of the liquid and of the solid, thus removing all restrictions of the above-described static methods, and would also take into account conformational equilibria in the liquid. No such calculations have yet been performed with the explicit aim of reproducing melting parameters.

As stated above, the detailed molecular mechanism of melting is not known. A relevant phenomenon, giving hints on this mystery, is surface melting. Surface molecules are less tightly bound than bulk molecules, and the amount of destabilization depends on surface structure; more 'open' surfaces, like high-Miller-indices surfaces in f.c.c. solids, are more easily disordered, and a thin film of liquid-like matter covers these surfaces well below the melting temperature. Remarkably, these liquid surface/solid bulk states are true thermodynamics equilibrium states. As temperature rises, the thickness of the film increases, and when the normal melting temperature is reached, the melt front rushes inwards and complete fusion is achieved. This phenomenon also gives a hint on the asymmetry of the melting/solidification process; while melting is quite naturally and inevitably initiated by the presence of surface liquid layers, solidification requires the formation of crystallization nuclei, which is kinetically controlled and can be more easily suppressed.

The above model has been experimentally validated for simple solids, like metals or noble gases, by spectroscopic, diffraction and even calorimetric methods. The fact that organic crystals do show the expected asymmetry in melting/solidification, even in the absence of more detailed experimental evidence, is a strong hint that surface melting is effective also for organic compounds. It is no wonder that a computational simulation of these phenomena is problematic.

2.3 The Solid–Solid Equilibrium

The formation of an organic crystal leads to a substantially rigid molecular assembly. Molecules are restricted in well-defined locations and orientations in space, as well as in a given molecular conformation, which may or may not be the absolute minimum-energy conformation for the isolated molecule. The free energy of each crystal structure is a function of temperature, and it turns out that a large majority of organic substances may appear in different crystal structures (polymorphs) in the relevant laboratory temperature range, say 273–400 K, at atmospheric pressure. Including the effects of pressure, the range of polymorphism is even wider.

The rigid crystal structures oppose a substantial activation energy towards disruption, so that transitions between solid phases are usually kinetically controlled. Moreover, they very seldom, if ever, proceed by a single-crystal to single-crystal mechanism. Partial or total decomposition of the starting structure, and growth of the daughter phase inside crystals of the mother one, usually lead to complex pathways whose final outcome is at best a polycrystalline sample of the final product. In many cases, one phase becomes metastable at a given temperature, but does not transform to the stable one because of the difficulty of breaking up the structure, and melting of the sample in its metastable form is observed. In other cases, the phase transition is indeed observed and melting of the metastable is not. Otherwise, both phases may separately reach equilibrium with the liquid, so that melting of both can be observed, but not the phase transition.

Coexistence of more than one crystal phase in the same crystal crop from the same solvent is due to nucleation-induced metastability, and is an infrequent, but not impossible, phenomenon. Growth from vapor or from melt is less likely to produce coexisting polymorphs.

Calorimetry is a straightforward method for detection of phase changes in solids. However, there are several experimental difficulties, since it is not always clear whether a given phase is stable or metastable. The case is made even more complicated by the appearance of transitions to disordered or plastic phases, or by other premelting phenomena. When two phases melt separately, the difference in the heats of melting, as measured from the area of the relevant DSC peaks, is equal to the enthalpy difference between the two solid phases, but the experimental uncertainty of these measurements can be very high.

The determination of enthalpy, entropy and free-energy differences between solid phases can be carried out through solubility measurements. The method is in principle very straightforward, but a number of experimental uncertainties affect the quality of the results. The dissolution equilibrium for a polymorph A can be formally written as

$$\text{A(crystal)} \dashrightarrow \text{A(solution)}$$

The equilibrium constant is the ratio of the activities:

$$K_{eq} = a(\text{A, sol})/a(\text{A, c}) = [\text{A, sol}] \tag{26}$$

being $a(A, \text{c}) = 1$ by definition, and $a(\text{A, sol}) = [\text{A, sol}]$ in dilute solutions, as is usually the case with organic solutes. Denoting the equilibrium solubility by Σ, the standard free-energy difference between the polymorphs can be obtained from the standard free energies of dissolution:

$$\Delta G^0_{\text{A}} = -RT \ln \Sigma(\text{A}); \quad \Delta G^{0\,'}_{\text{B}} = -RT \ln \Sigma(\text{B});$$
$$\Delta G^0_{\text{AB}} = \Delta G^0_{\text{A}} - \Delta G^0_{\text{B}} = -RT \ln (\Sigma(\text{A})/\Sigma(\text{B})) \tag{27}$$

At the thermodynamic transition temperature, the two phases are in equilibrium, so that $\Delta G^0_{\text{AB}} = 0$ and $\Sigma(\text{A}) = \Sigma(\text{B})$. When Σ is measured as a function of temperature, the enthalpy of dissolution, ΔH_{sol}, can be obtained from a van't Hoff plot, and the enthalpy difference between the two phases is obtained by Hess's law as $\Delta H_{\text{AB}} = \Delta H(\text{sol})_{\text{A}} - \Delta H(\text{sol})_{\text{B}}$. Once ΔG^0_{AB} and ΔH_{AB} are known, ΔS^0_{AB} can also be obtained.

A summary of recent experience in the study of polymorphism in organic crystals suggests that nearly all organic substances can crystallize in different crystal forms whose relative stability is a function of temperature (and pressure). Thus, studies at variable pressure and temperature almost inevitably reveal polymorphic phases. On the other hand, the appearance of different polymorphs at room conditions is a less frequent phenomenon. The way of saying (from a statement by McCrone) that the number of polymorphs for a given organic compound is proportional to the time and money spent in looking for them, should be completed by 'at variable temperature and pressure'.

True modeling of solid–solid equilibria requires a very accurate, separate evaluation of the free energy of each phase as a function of temperature, a goal that can be achieved only for molecular crystals made of very simple entities. For complex organic molecules, the calculation of the crystal energies and external entropies allows a straightforward estimation of sublimation parameters for each polymorph, and hence of differences between phases, when the crystal structures are known (see Section 2.1). At least for rigid molecules, the relative stability of two polymorphs can usually be judged using the calculated packing energy, since external entropy differences between phases are very small. Statistical studies of the properties of polymorphic crystals appearing at room temperature, as retrieved from crystallographic databases, also provide interesting clues.

2.4 Solid Solutions

The standard thermodynamic treatment of solutions between compounds A and B is in terms of enthalpies and entropies of mixing as a function of the mole fraction, x. For ideal solutions, the A...A, B...B and A...B potentials, and the molar volumes of A and B, are by definition the same, so that the enthalpy of mixing, ΔH_{AB}, is zero for any value of x. The entropy of mixing, ΔS_{AB}, is always positive (proceeding from unit mole fraction to a solution yields $\Delta G = RT \ln x$, or $\Delta S = -R \ln x$, positive since $x < 1$). Therefore, $\Delta G_{AB} = -T\Delta S_{AB}$ is always negative, and mixing is thermodynamically favored for any value of x.

For non-ideal solutions, ΔH_{AB} and ΔS_{AB} can take any value as a function of composition, and ΔG_{AB} may be positive or negative or even change sign at a given composition, thus defining a range of thermodynamic solubility. Excess thermodynamic functions are defined as the difference between the actual function and the ideal-solution function, so that H_{AB} (excess) $= \Delta H_{AB}$, and $G_{AB}(\text{excess}) = \Delta G_{AB} + T\Delta S_{AB}$. In the case of solid and liquid solutions, non-ideality can be treated in terms of a geometrical factor, related to molecular bulk and shape, and a molecular potential factor, related to the nature of the intermolecular forces involved. Coefficients of shape homeomorphism, related to geometry, can be defined rather easily from a consideration of the overlapping (V_O) and non-overlapping (V_{NO}) parts of the molecular volume, or as the difference between cell volumes, V_i:

$$\varepsilon_K = 1 - V_{NO}/V_O \tag{28}$$

$$\varepsilon_V = 1 - |(V_B - V_A)|/V_A \tag{29}$$

On the other hand, a measure of 'energetic' homeomorphism is, quite understandably, a matter of subjective appreciation of the various energy contributions.

The experimental analysis of such systems proceeds through measurements of the thermodynamic functions of mixing and of the parameters of the solid–liquid equilibrium, as well as, for the part concerned with geometry, through an analysis of the molecular homeomorphism coefficient from X-ray single-crystal studies of the separate components. From the computational side, an estimate of the enthalpy of mixing can be obtained as follows, if the crystal structures of A and B are known. Let PPE_A^* and PPE_B^* be the packing potential energies of the pure crystals. Then a simple calculation yields the variation in packing potential energy for the creation of a point defect occupied by A in the crystal of B (PPE(A in B)) and vice versa:

$$\text{PPE}(\text{mix}, x_A = 0) = \text{PPE}(\text{A in B}) - \text{PPE}_B^* \tag{30}$$

$$\text{PPE}(\text{mix}, x_B = 0) = \text{PPE}(\text{B in A}) - \text{PPE}_A^*$$

Table 3.5 Calculated potential packing energies (kcal/mole) and coefficients of homeomorphism[a] for solid solutions of 2X-naphthalenes in naphthalene

X	Molecular volume, Å^3	R(C-X), Å	PPE(B in A)	PPE(mix)	ε_K	ε_V
F	131	1.36	−33.9	−0.8	0.982	0.927
Cl	143	1.74	−31.2	1.9	0.893	0.839
Br	151	1.90	−16.6	16.5	0.847	0.806
I	159	2.10	12.0	45.1	—	—

[a]H. A. J. Oonk, P. R. van der Linde, Y. Haget, L. Bonpunt, N.-B. Chanh, M. A. Cuevas-Diarte, *J. Chim. Phys.*, **88**, 329 (1991)

Such a model neglects relaxation of the original structure at the point defect, and PPE(mix) approximates the enthalpy of mixing in the limit of infinite dilution, since a single-molecule defect has been created. A more realistic calculation of the true enthalpy of mixing at a given mole fraction should include a random distribution of defects (substitutions) and a full relaxation of the complete structure. However, even a crude quantity as PPE(mix) may be useful for a preliminary judgement of the relative compatibilities of molecules in organic solid solutions, being an energy-weighted estimation of the molecular homeomorphism indices. Table 3.5 gives some examples, consistent with the fact that solid solutions of iodonaphthalene in naphthalene are not observed.

2.5 Solvates and Inclusion Compounds

Quite often, the formation of an organic crystal from its solution requires (or allows for) the inclusion of a guest molecule in the crystal structure. Denoting as a 'guest' any substance present in the solution, except for impurities and the compound to be crystallized, a Cambridge Structural Database survey puts the occurrence of this phenomenon at 20% of the structures, with as many as 147 different cocrystallized guests (water in 61% of the cases). This, even more than the occurrence of polymorphism, is a severe obstacle to organic crystal-structure prediction. There is in fact no criterion for predicting, let alone controlling, the uptake of guest molecules by a given crystallizing compound. Rules based on molecular shape can lead to the prediction of possible uptake, like, for example, when the molecular structure has empty space between bulky groups. Nevertheless, in most cases these compounds are also capable of crystallizing in pure form, without uptake of guest molecules, so that possibility is never a necessity.

Guest molecules sometimes crystallize with site symmetry corresponding to their point-group symmetry, and/or on sites whose symmetry can be fulfilled by disorder (e.g. 50% occupancy of two orientations of toluene on an inversion center). The function of the guest molecule, in such cases, is that of space filler; crystallizing host molecules cannot get too close to crystal symmetry elements without repulsive steric overlap, and the empty space around the symmetry element is conveniently filled by guest molecules, either randomly taken from the solvent, or purposely supplied with the aim of host–guest cocrystallization. Guests with hydrogen-bonding ability have been found to intermesh in virtually any sort of hydrogen-bonding patterns with the crystallizing host.

One may wonder whether the uptake of guest molecules is decided at crystal nucleation and growth stages, or host–guest complex nuclei are already preformed in solution. The formation of a complex in solution could be monitored by spectroscopy, but its lifetime and concentration could be critical for spectroscopy but sufficient for crystal nucleation steering. The time and space resolution of complex formation in dilute solutions and at crystal nucleation stage is still a field open to speculation and future experiment.

Aside from X-ray diffraction, the most revealing experimental technique in host–guest crystal chemistry, when enough crystalline material is available, is the combination of thermogravimetry (to measure weight loss as a function of temperature) and calorimetry (to measure the amount of heat exchanged upon loss of guest molecules from the host–guest crystal). The heat of desorption of the guest can be, in principle, directly measured in that way; in practice, the host crystal phase often rearranges to its own uncomplexed crystal structure, instead of leaving behind an 'empty' solvate structure, so that the measured enthalpy is in fact the sum of two contributions.

From the computational point of view, the following considerations apply. Let PPE_{HG} be the packing potential energy of the host–guest crystal complex, and PPE_{H} that of the crystal structure of the pure uncomplexed host; both can be readily calculated if the crystal structures are known. Since the calculated energies are additive over atoms (if the atom–atom approach is taken) the following decomposition can be performed:

$$\begin{aligned} PPE_{HG} &= PPE(\text{host}-\text{host}) + PPE(\text{guest}-\text{guest}) + PPE(\text{host}-\text{guest}) \qquad (31) \\ &= PPE(\text{host}-\text{host}) - PPE(\text{des}) \end{aligned}$$

The first term is by definition the PPE of the 'empty' solvate structure. The potential energy loss on desorption, PPE(des), is the sum of the last two terms; PPE(guest–guest) is usually very small since guest molecules are normally very far from each other in the complex structure. The potential energy for rearrangement from the 'empty' solvate structure to the host's own uncomplexed crystal structure is then:

$$PPE(\text{rearr}) = PPE_{H} - PPE(\text{host}-\text{host}) \qquad (32)$$

Comparisons with experimentally measured heats of desorption should then include either PPE(des) alone, or the sum of PPE(des) and PPE(rearr), whichever is the case: X-ray work on the solid after desorption (even by powder methods) can usually ascertain whether rearrangement to the original uncomplexed crystal structure has occurred.

2.6 Summary

The device of atom–atom potentials, an energy calculation scheme that depends only on the packing geometry, bridges the gap between structural chemistry and thermodynamics. The preceding sections show how knowledge of the crystal structure allows the calculation of the lattice vibrations, and of thermodynamic quantities like the enthalpy, entropy and free energy of sublimation, the vapor pressure and the specific heat of the solid. Mechanical, optical and magnetic properties can also be calculated, given the molecular wavefunction. Simple energy calculations allow guesses at the enthalpy of melting, and at the potential-energy barriers for molecular reorientations in the solid, as well as an estimate of the heats of mixing and desorption in solid solutions and inclusion compounds. An X-ray crystal-structure determination, far from being an accomplished and final experiment, may just be the first step in a full investigation of solid-state chemical and physical properties.

3. KINETICS OF CRYSTAL DISSOLUTION

The dissolution of a crystal in an appropriate solvent can be monitored quite easily, by measuring the concentration of the resulting solution as a function of time; such measurements are routinely carried out on drug crystals, since solubility is an important parameter in drug action. The dissolution rate in principle carries information on the cohesion energy and on the structure of the crystal. Some principles and experimental aspects of crystal dissolution kinetics will be briefly reviewed in the following.

In the experimental setup, the solid phase can be in the form of a single crystal, dispersed particles, or a tablet; the solid is by definition at unit activity (not considering solvates) and supplies an infinite reservoir of substance. At infinite time, the saturated solution is in equilibrium with the solid; each polymorphic form has its own solubility, but solutions saturated with respect to metastable polymorphs are also metastable. Solution-mediated phase transformations are not uncommon. Temperature, pH, stirring speed and crystal morphology are other important parameters in the experiment. The dissolution rate is always written as $\mathrm{d}S/\mathrm{d}t$ and all kinetic equations are first-order, with possible reduction to zero order. The 'sink condition' is defined as the first stage of the dissolution, when the solute concentration is very small. Some relevant properties and symbols used in the treatment of dissolution kinetics are given in Table 3.6.

The main difficulty in the treatment of such a system is the application of

Table 3.6 Symbols and properties in dissolution kinetics

Symbol	Property	Units
$[A]$	available solid surface	$dm^2\ dm^{-3}$
ρ	density of the solid	$g{\cdot}dm^{-3}$
W_P	particle weight	g
D_P	crystal surface population	$mmole/dm^2$
S	solute concentration	mM
S^{eq}	equilibrium solute concentration	mM
h	thickness of the diffusion layer	dm
ε	diffusion coefficient	$time^{-1}\ dm^3\ dm^{-1}$
A, B, C . . .	solid phases in order of decreasing stability	
k_A	dissolution rate constant	$time^{-1}$
K_A	precipitation rate constant	$time^{-1}$
K_A^{NW}	apparent Noyes–Whitney constant, $k_A^{NW}{\cdot}[A]$	$time^{-1}$

kinetic equations for homogeneous solutions to what is an essentially heterogeneous system. Formally, the kinetic equation for a dissolution–precipitation equilibrium can be written as:

$$dS/dt = k_A A - K_A S \tag{33}$$

where A is formally the 'concentration' of the solid. While solids have by definition unit activity, an effective solid activity could be defined as $A = [A]{\cdot}D_P$, the product of an available surface with the population of molecules on that surface. Both quantities are rather difficult to quantify, but at least the kinetic equations become dimensionally correct. If the available area is constant, $k' = k_A A$ can be experimentally determined from the zero-order behavior in the sink condition, where $dS/dt = k'$. At equilibrium, $dS/dt = 0$, $K_A = k'/S_A^{eq}$; substituting and rearranging yields

$$S = S_A^{eq}[1 - \exp(-K_A t)] \tag{34}$$

In a different perspective, detachment of molecules from the crystal surface is considered to have a much higher rate constant than diffusion, and dissolution is considered to be diffusion-controlled. A saturated layer of thickness h is formed near the solid surface, and the kinetic model for diffusion from the saturated layer into bulk solution is the so-called Noyes–Whitney equation:

$$dS/dt = (\varepsilon[A]/h){\cdot}(S^{eq} - S) \tag{35}$$

which leads to the integrated form:

$$S = S_A^{eq}{\cdot}[1 - \exp(-K_A^{NW}{\cdot}t)]; \quad K_A^{NW} = [A]\varepsilon/h \tag{36}$$

This equation, in the sink condition, if often used in a form that represents the weight of dispersed particles as a function of time (the Hickson–Crowell equation):

$$(W_P^0)^{1/3} - W_P^{1/3} = (\tfrac{4}{3}\cdot\pi\cdot\rho)^{1/3}[(k_A^{NW}\cdot S_A^{eq}\cdot 10^3\cdot M)/\rho]\cdot t = K^{app}\cdot t \quad (37)$$

The integrated form of the Noyes–Whitney equation is formally identical to Eq. (34), and agreement of this equation with measured dissolution profiles, when no solution-mediated interconversion occurs, is usually good. The interpretation of the experimental apparent rate constants is, however, somewhat arbitrary.

If the detachment and diffusion stages are considered to have comparable rate constants, the measured, apparent rate constant in Eq. (36) can be partitioned as follows:

$$K_{app} = k_r k_D/(k_r + k_D) \quad (38)$$

where k_r is the rate constant for the surface reaction supplying molecules to the high concentration layer, and k_D is the diffusion constant. k_r should carry information on attachment energies within crystals.

The kinetic treatment of solution-mediated phase transformations is rather complicated. The dissolution profile shows a maximum, corresponding to competition between the solubility of the two crystal forms, and then decays to the solubility of the more stable (less soluble) polymorph. A simplified treatment can be proposed as follows, under the conditions that (a) solid B, the less stable polymorph, is finely dispersed at the beginning of the dissolution; (b) there is no direct B-to-A solid–solid transformation; (c) the active concentration of the solids is proportional to the total mass, as in finely dispersed particle systems (thus minimizing also the effects of morphology); (d) stirring is fast, so that diffusion is almost instantaneous. These conditions represent the best possible approximation to a homogeneous system. A and B are then the active concentrations of solid.

The treatment can be simplified by assuming that recrystallization towards the less stable polymorph can be neglected:

$$\text{solid B} \xleftarrow{k_B} S \underset{k_A}{\overset{K_A}{\rightleftarrows}} \text{solid A}$$

The mathematical developments are:

$$-\mathrm{d}B/\mathrm{d}t = k_B\cdot B; \quad B = B^0\exp(-k_B\cdot t) \quad (B^0 = A + B + S) \quad (39)$$

$$\begin{aligned}\mathrm{d}S/\mathrm{d}t &= k_B\cdot B + k_A\cdot A - K_A\cdot S\\ &= k_B\cdot B^0\cdot\exp(-k_B\cdot t) + k_A[B^0 - B^0\cdot\exp(-k_B t) - S] - K_A\cdot S \quad (40)\end{aligned}$$

Defining the following auxiliary variables and integrating:

$$P = k_A\cdot B^0; \quad Q = B^0(k_B - k_A); \quad T = (k_A + K_A);$$

$$U = Q/(T - k_B); \quad Z = P/T;$$

$$S = U\cdot\exp(-k_B\cdot t) - (U + Z)\cdot\exp(-T\cdot t) + Z \quad (41)$$

The boundary conditions are:

(a) $$t = 0\ ,\ S = 0; \quad t = \infty\ ,\ S = S_{\mathrm{A}}^{\mathrm{eq}} = Z = k_{\mathrm{A}} \cdot B^0/(k_{\mathrm{A}} + K_{\mathrm{A}}); \quad (42)$$

(b) $\mathrm{d}S/\mathrm{d}t = 0$ for $t = t^*$ (the time of the maximum concentration of S):

$$S^* = (Q/T) \cdot \exp(-k_{\mathrm{B}} \cdot t^*) + S_{\mathrm{A}}^{\mathrm{eq}}. \quad (43)$$

Since $v^0 = k_{\mathrm{B}} \cdot B^0$,

$$S^* - S_{\mathrm{A}}^{\mathrm{eq}} = (v^0 - B^0 \cdot k_{\mathrm{A}})/(k_{\mathrm{A}} + K_{\mathrm{A}}) \exp(-k_{\mathrm{B}} \cdot t^*) \quad (44)$$

Substituting and rearranging:

$$k_{\mathrm{A}} + K_{\mathrm{A}} = v^0 \cdot W/Y \quad (43)$$

$$W = \exp(-k_{\mathrm{B}} \cdot t^*); \quad Y = S^* + [\exp(-k_{\mathrm{B}} \cdot t^*) - 1] \cdot S_{\mathrm{A}}^{\mathrm{eq}} \quad (44)$$

Since B^0, v^0, S^*, $S_{\mathrm{A}}^{\mathrm{eq}}$ and t^* are either known or measurable from the dissolution profiles, the kinetic system can be completely determined, calculating k_{B} as v^0/B^0, and then $k_{\mathrm{A}} + K_{\mathrm{A}}$; substitution into (42) gives k_{A}.

Computer modeling of dissolution kinetics at a molecular level is still well beyond reach at present. The theoretical treatment of dissolution relies on phenomenological equations, as in the above examples.

4. CRYSTAL-STRUCTURE PREDICTION AND CONTROL

This section addresses the problem of predicting and controlling some or all of the crystal properties of a given organic compound, by computer simulation (see also Chapters 1, 4 and 6). It will be assumed that the chemical composition and the connectivity of the molecule are known, while one or many possible molecular conformations may or may not have been assigned, either by experiment or computation; therefore a definite molecular model, in the form of coordinates for the atomic nuclei for input to computations, may or may not be available. Additional physicochemical, spectroscopic or crystallographic information may or may not be available from separate experiments or from chemical intuition.

The most complete and ambitious goal is prediction of the thermodynamically and kinetically allowed crystal structure (cell parameters, space group and atomic coordinates) from the bare chemical constitution of the molecule, at any given conditions. This requires accurate knowledge of the free energies of many polymorphs, as a function of temperature, including conformational flexibility; and of the kinetic factors (nucleation and growth as a function of temperature, nature of the solvent, possible impurities, etc.). This last point will not be discussed further, since there are at present no computational techniques for dealing with such complex systems. Some molecular models could be envisaged for growth from vapor or melt, but a dynamical, molecular-level simulation of growth from solution is at present impossible.

What can be accomplished rather easily is the computer generation of many (a great many) crystal structures and the calculation of the packing energy and external vibrational entropy of each of them. In such an approach, only empirical potential-energy schemes can be applied, because of the huge computational problem. Packing-energy differences should approximate enthalpy differences between phases. Entropy differences are usually small, so that relative stability can be judged on the basis of ΔH_{AB} only. These calculation are by definition temperature-independent, so that a real treatment of the temperature dependence of enthalpies and entropies of transition is impossible.

The basic (but naive) principle in deciding which structure, among the hundreds that are generated, will be the one actually appearing is to choose the most stable one, on the basis of calculated packing energies. The main obstacles against good predictive ability of this approach are:

(a) incomplete scanning of the very large crystal potential hypersurface. Even if the search is restricted to a few common space groups, the degrees of freedom are many: cell parameters, location and orientation of the molecule, internal conformational flexibility;
(b) inadequacies in the intermolecular force field. Energy differences between polymorphs are in the range of a few kJ/mole, and deciding which is the most stable depends on fine detail in the force fields. If the molecule is chemically simple, without interaction between several functional groups, then force fields optimized for single classes of compounds have more chances of being adequate;
(c) the lack of consistency in the treatment of intra- and intermolecular parts of the force field. When conformational freedom is present, summing energy contributions from separately optimized force fields is problematic.

Incidentally, even if the true crystal structure cannot be selected from the many generated ones, the enthalpy of sublimation of the crystal can be estimated, since the packing energy of any of the most stable computational polymorphs will be within 10 percentage points of the true sublimation enthalpy (see Eq. (6)).

When hydrogen bonding is present, chemical reasoning can be helpful in the assignment of some predictable recognition patterns, thus helping climb the first step of the crystal-structure prediction ladder; however, the prediction of the packing of hydrogen-bonded oligomers runs into the same difficulty as the prediction of the crystal packing of isolated molecules. For example, a carboxylic acid will almost certainly form a cyclic dimer in the crystal, but the packing of the dimer is far from predictable. For 1,3,5-trinitro-triaminobenzene, a pseudo-hexagonal layer with N—0. . .H—N hydrogen bonds is easily predicted, but then the further (and crucial, for optical properties) problem of predicting whether successive layers will arrange in a parallel or antiparallel fashion cannot be solved.

Most frequently, several steric or electronic requirements, from chemical functions or steering groups in polyfunctional molecules, compete in driving the intermolecular arrangement. Chemical intuition alone is unable to weigh properly the subtle balance among them; a quantitative energetic treatment is as imperative as it is problematic.

On the other hand, even fragmentary or incomplete crystallographic information can quickly lead to an unequivocal crystal structure prediction. The relative energies of competing minima depend on the force field, but the location of these minima on the hypersurface changes very little when the force field is changed (within reasonable limits). If some or all of the cell parameters, or the space group, are available even in the absence of diffraction intensity data, the present crystal-structure simulation procedures can reconstruct the whole crystal structure in short computing times, if the molecular structure is known. Whether the energy minimum that corresponds to the known cell parameters is the lowest energy one, is no longer relevant.

The production of single crystals suitable for X-ray diffraction is nowadays less critical, since modern equipment can use small or even unstable crystals, diffracting power and detection sensitivity having greatly increased. In spite of this, the production of single crystals suitable for a complete collection of intensity data can still be sometimes a problem; in such cases, a quick determination of cell parameters and space group could provide the necessary input to computational procedures for complete structure solution. If microcrystalline material is available, then high-resolution powder spectra can be collected and used as a fingerprint for checking the calculated crystal structure; refinement can then be accomplished by Rietveld analysis, nowadays implemented for crystal and molecular structure refinement even for medium-to-large organic molecules. It can be reasonably stated that in the near future the need for well-grown single crystals and for the collection of extensive X-ray diffraction intensity data for small to medium-size organic molecules may be greatly reduced.

5. SUGGESTIONS FOR FURTHER READING

5.1 Sources of Crystallographic Data in Machine-Readable Form, and Statistical Studies

The standard source of crystallographic data for organic and organometallic crystals is the Cambridge Structural Database (CSD); a good introduction to its structure and performance can be found in F. H. Allen, O. Kennard and D. G. Watson, 'Crystallographic Databases: Search and Retrieval and Information from the Cambridge Structural Database', in *Structure Correlation*, edited by H.-B. Bürgi and J. D. Dunitz, VCH, Weinheim (1994). The Database contains cell and space group data, and nuclear coordinates for all atoms in the basic chemical unit (a full molecule even when the asymmetric unit is a fraction of molecule). The accompanying software allows all sorts of screening for desired bonding arrangements, as well as for chemical functionality.

Once again, it is emphasized that hydrogen atom coordinates from X-ray work are not suitable for intermolecular studies. Renormalization is mandatory; see, for example, A Gavezzotti and G. Filippini, *J. Phys. Chem.*, **98**, 4831 (1994), and references therein.

The statistical package provided by the CSD corporation allows the statistical study of geometrical intra- and intermolecular properties; see F. H. Allen, J. E. Davies, J. J. Galloy, O. Johnson, O. Kennard, C. F. Macrae, E. M. Mitchell, G. F. Mitchell, J. M. Smith and D. G. Watson, *J. Chem. Inf. Comput. Sci.*, **31**, 187 (1991). The literature on statistical studies of the geometry of non-bonded interaction is rich and growing; see, for example, the references and discussions in G. R. Desiraju, *Crystal Engineering*, Elsevier, Amsterdam (1989), Chapters 5–8. Some more recent examples: B. Pirard, G. Baudoux and F. Durant, *Acta Cryst.*, **B51**, 103 (1995), for NH. . .O hydrogen bonds in carboxylates, sulfonates and monohydrogen phosphates; T. Burling and B. M. Goldstein, *Acta Cryst.*, **B49**, 738 (1993), for intramolecular sulfur–nucleophile contacts; P. Gilli, V. Bertolasi, V. Ferretti and G. Gilli, *J. Am. Chem. Soc.*, **116**, 909 (1994) for strong O—H. . .O hydrogen bonds; Th. Steiner and W. Saenger, *Acta Cryst.*, **B48**, 819 (1992), on O—H. . .O hydrogen bonds in carbohydrates.

A critical review of structure correlation methods is in V. Ferretti, P. Gilli, V. Bertolasi and G. Gilli, *Crystallography Reviews*, **5**, 1 (1996).

Coordinate manipulations on non-orthogonal reference systems are described in J. D. Dunitz, *X-Ray Analysis and the Structure of Organic Molecules*, Cornell University Press, Ithaca, NY (1979), introductory chapters. See also several chapters in *Structure Correlation*, edited by H.-B. Bürgi and J. D. Dunitz, VCH, Weinheim (1994): Chapter 1, H.-B. Bürgi and J. D. Dunitz, 'Molecular Structure and Coordinate Systems'; Chapter 12, A. Gavezzotti, 'Molecular Packing and Correlations between Molecular and Crystal Properties', Appendix.

5.2 Periodic Orbitals

A terse, chemically oriented description of the theory underlying the molecular periodic orbital method and electronic band theory is in R. Hoffmann, *Solids and Surfaces, A Chemist's View of Bonding in Extended Structures*, VCH, Weinheim (1988). The introductory chapters of this book reconcile the languages or solid-state physics and of periodic orbital quantum chemistry.

5.3 Lattice Vibrations

A detailed mathematical treatment of the dynamics of harmonic oscillations of molecules in crystals can be found, for example, in A. I. Kitaigorodski,

Molecular Crystals and Molecules, Academic Press, New York (1973), Chapter III, pp. 193–212, and in S. Califano, V. Schettino and N. Neto, *Lattice Dynamics of Molecular Crystals*, Springer-Verlag, Berlin (1981). Applications, with a more condensed but perhaps clearer theoretical description, can be found in G. Taddei, H. Bonadeo, M. P. Marzocchi and S. Califano, *J. Chem. Phys.*, **58**, 1966 (1973), and in G. Filippini, C. M. Gramaccioli, M. Simonetta and G. B. Suffritti, *J. Chem. Phys.*, **59**, 5088 (1973); G. Filippini, C. M. Gramaccioli, M. Simonetta and G. B. Suffritti, *Chem. Phys.*, **8**, 136 (1975) describe also in detail the statistical-mechanical derivation of crystal thermodynamic functions from vibrations data. Both papers refer to rigid hydrocarbons, while C. M. Gramaccioli and G. Filippini, *Acta Cryst.*, **A39**, 784 (1983) extend the procedure to include internal modes, with possible mixing with external (intermolecular) modes.

The molecular motions associated with a particular lattice vibrational mode can be visualized from the corresponding eigenvectors. Piezomodulated Raman spectroscopy can directly test these assignments (K. M. White and C. J. Eckhardt, *J. Chem. Phys.*, **92**, 2214 (1990)). For a combined X-ray, spectroscopic and lattice dynamical investigation of a reaction mechanism, see J. J. Stezowski, N. M. Peachey, P. Goebel and C. J. Eckhardt, *J. Am. Chem. Soc.*, **115**, 6499 (1993).

5.4 Mechanical Properties

An early attempt to calculate elastic properties of molecular crystals by atom–atom potentials is in K. V. Mirskaya, *Soviet Phys.-Crystallography*, **17**, 51 (1972); more recently, see P. Pavlides, D. Pugh and K. J. Roberts, *Acta Cryst.*, **A47**, 846 (1991). A paper by W. R. Busing and M. Matsui, *Acta Cryst.*, **A40**, 532 (1984) describes a method for the computational application of external forces (pressure, stresses, electrical fields) to organic crystal simulations. Generally speaking, however, the related literature is more akin to solid-state physics than solid-state chemistry.

5.5 Dielectric and Optical Properties

The interaction between molecular dipoles, taking into account mutual polarization, can be used to model a number of crystal properties, including the spectral intensities of lattice vibrational modes. The basic approach has been called 'microelectrostatics', meaning the use of standard electrostatics at the molecular level, dispensing with an explicit quantum-mechanical treatment. Molecules are characterized by an effective polarizability in the crystal environment, which can be calculated quantum-mechanically or deduced from experimental quantities. In particular, when the dipole moment induced in a molecule is not proportional to the polarizing field, a non-linear response, much sought after for high-technology applications in signal processing, is observed. See R. W. Munn, *Mol. Phys.*, **64**, 1 (1988).

At a macroscopic level, the study of optically anomalous crystals (whose optical symmetry is lower than would be expected from their external form or X-ray diffraction data) by the polarizing microscope reveals that impurities are incorporated by kinetically controlled crystal growth. See B. Kahr and J. M. McBride, *Angew. Chem. Int. Ed. Engl.*, **31**, 1 (1992).

5.6 Defects and Disorder

The following literature deals with those aspects of the subject that are amenable to computer modeling using simple potentials.

For a discussion of models of defects in crystals see A. J. Pertsin and A. I. Kitaigorodski, *The Atom–Atom Potential Method, Applications to Organic Molecular Solids*, Springer-Verlag, Berlin (1987), Chapter 7. Modeling of orientational disorder by a parametric equation using empirical atom–atom potentials is described by K. Rohleder, T. Luty and B. Kuchta, *J. Chem. Phys.*, **15**, 1573 (1994).

For experimental measurements (mostly by variable-T NMR) and calculations on the activation energies of molecular reorientations is solids by empirical potentials, see: D. Braga, *Acc. Chem. Res.*, **92**, 633 (1992); A. Gavezzotti and M. Simonetta, *Chem. Revs.*, **82**, 1, (1982); A. Gavezzotti and M. Simonetta, 'Molecular Motions in Organic Crystals: the Structural Point of View', in: *Organic Solid State Chemistry*, edited by G. R. Desiraju, Elsevier, Amsterdam (1987).

5.7 Solid–State Reactivity

See: *Organic Chemistry in Anisotropic Media, Tetrahedron Symposia-in-Print no. 29*, edited by V. Ramamurthy, J. R. Scheffer and N. J. Turro, *Tetrahedron*, vol. 43, pp. 1197–1745 (1987); N. B. Singh, R. J. Singh and N. P. Singh, Tetrahedron 50, 6441 (1994), for a wide presentation of experiments and calculations; Y. Ohashi, *Reactivity in Molecular Crystals*, VCH, Weinheim (1993); S. R. Byrn, *Solid-State Chemistry of Drugs*, Academic Press, New York (1982). For an explicit example of modeling the displacement and relaxation of reaction products in a crystalline matrix, see S. K. Kearsley and J. M. McBride, *Mol. Cryst. Liq. Cryst.*, **156**, 109 (1988), on the photolysis of acetyl benzoyl peroxide. In the related experiments, products are formed in such a small yield that they can be treated as point defects in the crystalline matrix. See also J. M. McBride, B. E. Segmuller, M. D. Hollingsworth, D. E. Mills and B. A. Weber, *Science*, **234**, 830 (1986). A review paper on phase changes and chemical reactions in molecular crystals, on both experimental studies and computer simulations, is J. D. Dunitz, *Acta Cryst.*, **B51**, 619 (1995).

The phase interconversions in *p*-(trimethylammonio)benzenesulfonate have been studied by X-ray crystallography and computer simulations: J. A. R. P.

Sarma and J. D. Dunitz, *Acta Cryst.*, **B46**, 784 (1990). Reaction paths for the solid-state racemization in crystalline cobaloxime complexes have been calculated by A. Uchida and J. D. Dunitz, *Acta Cryst.*, **B46**, 45 (1990).

5.8 Empirical Force Fields

The seminal work by D. E. Williams and co-workers on the optimization of crystal force fields has been reviewed extensively; only a few entry points to the relevant literature will therefore be given: D. E. Williams, 'Net Atomic Charge and Multipole Models for the ab Initio Molecular Electric Potential', in *Reviews in Computational Chemistry*, edited by K. B. Lipkowitz and D. B. Boyd, Vol. 2, Ch. 6, VCH, New York (1991); D. E. Williams and T. R. Stouch, *J. Comput. Chem.*, **14**, 1066 (1993).

A standard reference for empirical potentials, their nature, derivation and applications is A. J. Pertsin and A. I. Kitaigorodski, *The Atom–Atom Potential Method, Applications to Organic Molecular Solids*, Springer-Verlag, Berlin (1987). The book also carries a theoretical treatment of lattice-dynamical and thermodynamic calculations in organic crystals.

Several contributions from the authors' group discuss in detail the relative merits of various empirical formulations of crystal potentials: see G. Filippini and A. Gavezzotti, *Acta Cryst.*, **B49**, 868 (1993); A. Gavezzotti and G. Filippini, *J. Phys. Chem.*, **98**, 4831 (1994); and references therein.

It was early realized (A. Warshel and S. Lifson, *J. Chem. Phys.*, **33**, 582 (1970)) that only a consistent, simultaneous optimization of intra- and intermolecular force fields can give good results. For recent examples of such a derivation, for biologically relevant molecules, see A. D. MacKerell, J. Wiorkiewicz-Kuczera and M. Karplus, *J. Am. Chem. Soc.*, **117**, 11946 (1995); W. D. Cornell, P. Cieplak, C. I. Bayly, I. R. Gould, K. M. Merz, D. M. Ferguson, D. C. Spellmeyer, T. Fox, J. W. Caldwell and P. A. Kollman, *J. Am. Chem. Soc.*, **117**, 5179 (1995). Force fields for general application in molecular dynamics and Monte Carlo calculations on liquids and solutions were developed earlier; see W. H. van Gunsteren and H. J. C. Berendsen, *Angew. Chem. Int. Ed. Engl.*, **29**, 992 (1990); P. A. Kollman and K. M. Merz, *Acc. Chem. Res.*, **23**, 246 (1990); W. L. Jorgensen, *Chemtracts-Org. Chemistry*, **4**, 91 (1991).

5.9 Quantum-Chemical Calculations

For an *ab initio* and semi-empirical molecular orbital calculation on aggregates containing up to 36 molecules of acetic acid, see L. Turi and J. J. Dannenberg, *J. Am. Chem. Soc.*, **116**, 8714 (1994). For periodic orbital calculations on an organic crystal, including electron correlation at the MP4 level, see S. Suhai, *J. Chem. Phys.*, **103**, 7030 (1995). Not surprisingly, the substance considered is a six-atom one, formamide.

An *ab initio* crystal field method, by which one or a few molecules are treated quantum-chemically, and the crystal environment is simulated by an electrostatic field alone, has been developed; see its application to amino acids in A. Peeters, C. Van Alsenoy, A. T. H. Lenstra and H. J. Geise, *J. Chem. Phys.*, **103**, 6608 (1995). Admittedly, the simulation of the crystal field is intrinsically poor in what concerns intermolecular interactions (see the Conclusions section of the above paper). These methods probably describe in an acceptable way stronger local interactions (like hydrogen bonds), but fail to describe the weak intermolecular forces which are crucial in determining the crystal structures of organic compounds.

For *ab initio* calculations on the intermolecular potential energy surface for small-sample systems (water ... acetylene, water ... ethylene, water ... methane) see M. C. Rovira, J. J. Novoa, M.-H. Whangbo and J. M. Williams, *Chem. Phys.*, **200**, 319 (1995). Sorting out relevant interactions in crystals in such a way is sometimes dangerous, but is a promising step towards a quantitative analysis of main contacts in a crystallographic context; see, for example, a study of C—H...S and S...S in crystals of organic conductors, in J. J. Novoa, M. C. Rovira, C. Rovira, J. Veciana and J. Tarres, *Adv. Mat.*, **7**, 233 (1995). One serious drawback in such studies is the basis-set superposition error (BSSE), which is usually corrected by the counterpoise method. An alternative formulation, which cuts the problem at its origin by expanding atomic orbitals on separate fragments only, is in E. Gianinetti, M. Raimondi and E. Tornaghi, *Int. J. Quantum Chem.* **60**, 157 (1996).

5.10 General Thermodynamics

A clear and comprehensive introduction to chemical thermodynamics, in both its classical and its statistical foundations, is D. R. Stull, E. F. Westrum, Jr., and G. C. Sinke, *The Chemical Thermodynamics of Organic Compounds*, John Wiley and Sons, New York (1969). Another fundamental reading is E. F. Westrum, Jr. and J. P. McCullough, 'Thermodynamics of Crystals', in *Physics and Chemistry of the Organic Solid State*, edited by D. Fox, M. M. Labes and A. Weissberger, Vol. 1, Wiley-Interscience, New York (1963).

A standard reference for structural and thermodynamic properties of condensed phases is A. Bondi, *Physical Properties of Molecular Crystals, Liquids and Glasses*, John Wiley and Sons, New York (1968). This contribution, although old, provides an indispensable framework for an understanding of general principles.

Solid–vapor equilibrium: a seminal paper on the measurement and calculation of heats of sublimation, by a scheme using group increments, is A. Bondi, *J. Chem. Eng. Data*, **8**, 371 (1963).

For a review of available sublimation enthalpies of organic compounds see

J. S. Chickos, in *Molecular Structure and Energetics*, Vol. 2, edited by J. F. Liebman and A. Greenberg, VCH, New York (1987).

Solid–liquid equilibrium: a rather old but still fundamental contribution is A. R. Ubbelohde, *The Molten State of Matter*, John Wiley and Sons, Chichester (1978). This book carries the subtitle 'Melting and Crystal Structure', and somehow reflects on initial enthusiasm, spurred by advances in X-ray diffraction crystallography, for the possibility of explaining melting from crystal structural data. See, in particular, pp. 12, 13, 131, 148 for volume variations on melting and pp. 124 ff. for positional melting.

For a compilation of melting temperatures and melting enthalpies of organic compounds, see W. E. Acree, *Thermochim. Acta*, **189**, 37 (1991) and **219**, 97 (1993); J. S. Chickos, C. M. Braton, D. G. Hesse and J. F. Liebman, *J. Org. Chem.* **56**, 927 (1991).

On the irreversible nature of the fusion/solidification equilibrium see, for example, C. A. Angell, L. E. Busse, E. I. Cooper, R. K. Kadiyala, A. Dworkin, M. Ghelfenstein, H. Szwarc and A. Vassal, *J. Chim. Phys.*, **82**, 267 (1985); these authors state that 'in a randomly chosen group of organic chemicals . . . nearly half of those which can be fused without decomposition will fail to crystallize during fast cooling'. On relationships between molecular symmetry and melting characteristics see A. Gavezzotti, *J. Chem. Soc. Perkin Trans.*, **2**, 1399 (1995).

Melting is much better studied in inorganic crystals. See, for example, M. P. Tosi, D. L. Price and M.-L. Saboungi, 'Ordering in Metal Halide Melts', in *Ann. Rev. Phys. Chem.*, edited by H. L. Strauss, G. T. Babcock and S. R. Leone, vol. 44, p. 173 (1993). Modern analytical techniques are beginning to be used for direct observation of phase transitions in solids: see e.g. H. Dai and C. M. Lieber, 'Scanning Tunneling Microscopy Studies of Low-Dimensional Materials: Charge Density Wave Pinning and Melting in Two Dimensions', in *Ann. Rev. Phys. Chem.*, edited by H. L. Strauss, G. T. Babcock and S. R. Leone, vol. 44, p. 237 (1993).

Surface melting is clearly described in a paper by J. W. M. Frenken, *Endeavour*, **14**, 2 (1990). Unfortunately for organic chemists, as the author states, investigators of this phenomenon had to turn 'to much simpler solids than water, namely those consisting of only one type of atom'.

For an embryo of theories of molecular melting, see M. Keskin and S. Ozgan, *Mol. Cryst. Liq. Cryst.*, **269**, 149 (1995); the references in this paper also refer to earlier theories, mostly based on structureless 'molecules'.

5.11 Solid–Solid Equilibria and Polymorphism

For a calculation of the phase diagram of a simple crystal (solid nitrogen) see G. Filippini, C. M. Gramaccioli, M. Simonetta and G. B. Suffritti, *Mol. Phys.*, **35**, 1659 (1978). For benzene, see R. G. Della Valle, E. Venuti and A. Brillante, *Chem. Phys.*, **202**, 231 (1996).

Polymorphism in organic crystals has recently been reviewed, from a thermodynamic, structural and computational perspective, by A. Gavezzotti and G. Filippini, *J. Am. Chem. Soc.*, **117**, 12299 (1995); references in that paper cover most of the relevant literature, and will not be repeated here. Statistical correlations were also established using the polymorphic structures determined by X-ray diffraction. 'Disappearing Polymorphs' is the appealing title of a paper by J. D. Dunitz and J. Bernstein, *Acc. Chem. Res.*, **28**, 193 (1995). For a review of the subject from a different perspective, see: 'Analysis of Organic Polymorphs', by T. L. Threlfall, *Analyst*, **120**, 2435 (1995).

Polymorphism is of increasing concern to the pharmaceutical industry for its relevance to solid-state formulations. Since studying polymorphism and predicting crystal structures are two facets of the same topic, see also the section on crystal-structure prediction in this chapter.

5.12 Solid Solutions

A seminal contribution is A. I. Kitaigorodski, *Mixed Crystals*, Springer-Verlag, Berlin (1984). Results on naphthalene and benzene derivatives, including a cooperation between X-ray and thermodynamical measurements, are given in H. A. J. Oonk, P. R. van der Linde, Y. Haget, L. Bonpunt, N.-B. Chanh and M. A. Cuevas-Diarte, *J. Chim. Phys.*, **88**, 329 (1991); L. Bonpunt, R. Courchinoux, Y. Haget, E. Estop, T. Calvet, M. A. Cuevas-Diarte and M. Labrador, *J. Appl. Cryst.*, **24**, 164 (1991).

5.13 Solvates

P. van der Sluis and J. Kroon, *J. Cryst. Growth*, **97**, 645 (1989) reviewed the subject using the Cambridge Structural Database.

A vast literature deals with the structural and energetic aspects of the formation of solid solvates, clathrates and inclusion compounds. See, for example, E. Weber, 'Molecular Inclusion and Molecular Recognition', in *Topics in Current Chemistry 140* (1987) and *149* (1988), Springer-Verlag, Berlin-Heidelberg; *Inclusion Compounds*, edited by J. L. Atwood, J. E. D. Davies and D. D. MacNicol, Vol. 1–3, Academic Press, London (1984). An example of simultaneous application of X-ray crystallography and thermodynamical measurements is S. A. Bourne, L. R. Nassimbeni and F. Toda, *J. Chem. Soc. Perkin Trans.*, **2**, 1335 (1991). For an example of the mechanical and optical versatility of urea inclusion compounds, see M. E. Brown and M. D. Hollingsworth, *Nature*, **376**, 323 (1995).

For a computational simulation of channels and voids in inclusion compounds, see the volume-packing analysis method in A. Gavezzotti, *J. Am. Chem. Soc.*, **105**, 5220 (1983), and applications, for example, in M. H. Moore, L. R. Nassimbeni and M. L. Niven, *Inorganica Chim. Acta*, **131**, 45 (1987).

Crystal-packing energy calculations have been used to study the structure and energetics of guest molecules in channels: R. Popovitz-Biro, C. P. Tang, H. C. Chang, M. Lahav and L. Leiserowitz, *J. Am. Chem. Soc.,* **107**, 4043 (1985); S. Candeloro De Sanctis, V. M. Coiro, F. Mazza and G. Pochetti, *Acta Cryst.*, **B51**, 81 (1995).

5.14 Dissolution Kinetics

Early (but still relevant) papers on the theory of dissolution kinetics are: A. A. Noyes and W. R. Whitney, *J. Am. Chem. Soc.*, **19**, 930 (1897); A. Berthoud, *J. Chim. Phys.*, **10**, 624 (1912). For the separation of kinetic constants for detachment from the solid and diffusion, see D. E. Wurster and P. W. Taylor, *J. Pharm. Sci.*, **54**, 670 (1965). For an example of more advanced techniques for the simulation of dissolution kinetics in particle systems, see Y. Yonezawa, I. Shinohara, A. Otsuka and H. Sundada, *Chem. Pharm. Bull.*, **38**, 1024 (1990) and references therein. A vast literature reports on experimental determinations of dissolution curves of pharmaceuticals.

For modern views on probing crystal-dissolution kinetics at the molecular level, see P. R. Unwin and J. V. Macpherson, *Chem. Soc. Revs.*, **24**, 109 (1995). Atomic force microscopy has been used to study *in situ* the dissolution of organic crystals (1,2-diphenylethane derivatives); J. A. Swift and J. M. McBride, *NATO ARW Crystals: Supramolecular Materials*, Sestri Levante (1995), abstracts book and personal communication to the authors of this chapter.

5.15 Crystal-Structure Prediction and Control

For the use of chemical intuition in crystal-structure prediction, and a rationalization of hydrogen-bonding patterns, see L. Leiserowitz and A. T. Hagler, *Proc. Roy. Soc. Lond.*, **A388**, 133 (1983); J. Bernstein, M. C. Etter and L. Leiserowitz, in *Structure Correlation*, edited by H.-B. Bürgi and J. D. Dunitz, VCH, Weinheim (1994), Chapter 11. See also C. B. Aakeroy and K. R. Seddon, *Chem. Soc. Revs.*, **22**, 397 (1993); J. H. Williams, *Acc. Chem. Res.*, **26**, 593 (1993); A. D. Burrows, C.-W. Chan, M. M. Chowdhry, J. E. McGrady and D. M. P. Mingos, *Chem. Soc. Revs.*, **24**, 329 (1995); G. R. Desiraju, *Angew. Chem. Int. Ed. Engl.*, **34**, 2311 (1995). This last paper proposes an analogy between molecular synthesis and crystal 'synthesis'.

The following is a list of computational approaches to the problem of crystal-structure determination: D. Rabinovitch and G. M. J. Schmidt, *Nature*, **211**, 1391 (1966); D. Rabinovich and Z. Shakked, *Acta Cryst.*, **A40**, 195 (1984); S. Perez, C. Vergelati and V. H. Tran, *Acta Cryst.*, **B41**, 262 (1985); H. L. Ammon, Z. Du, J. R. Holden and L. A. Paquette, *Acta Cryst.*, **B50**, 216 (1994); J. R. Holden, Z. Du and H. L. Ammon, *J. Comput. Chem.*, **14**, 422 (1993) (program MOLPAK); H. R. Karfunkel and R. J. Gdanitz, *J. Comput. Chem.*, **13**, 1171 (1992); W. Linert and F. Renz, *J. Chem. Inf. Comput. Sci.*, **33**, 776 (1993); B. P. van Eijck, W. T. M. Mooij and J. Kroon, *Acta Cryst.*, **B51**, 99 (1995); T. Shoda and D. E. Williams, *J. Mol. Struct.* THEOCHEM, **357**, 1 (1995) (MPA-MPG package); M. U. Schmidt and U. Englert, *J. Chem. Soc.* Dalton 2077 (1996); A. Gavezzotti, *Acta Cryst.*, **B52**, 201 (1996); A. M. Chaka, R. Zaniewski, W. Youngs, C. Tessier and G. Klopman, *Acta Cryst.*, **B52**, 165 (1996) (ICEG program).

For the simulation and prediction of linear arrays or monolayers, see R. P. Scaringe and S. Perez, *J. Phys. Chem.*, **91**, 2394 (1987); A. Ulman and R. P. Scaringe, *Langmuir*, **8**, 894 (1992); J. Perlstein, *J. Am. Chem. Soc.*, **116**, 11420 (1994).

For recent developments in minimum-finding and energy-surface search algorithms, see K. D. Gibson and H. A. Scheraga, *J. Phys. Chem.*, **99**, 3765 (1995); X. Xiao and D. E. Williams, *Chem. Phys. Letters*, **215**, 17 (1993); D. E. Williams, *Chem. Phys. Letters*, **192**, 538 (1992).

6. APPENDIX: ANALYTICAL FORMS OF EMPIRICAL ATOM–ATOM FORCE FIELDS

The basic parameters in an atom–atom interaction energy curve are the position of the minimum, R, the well depth, ε, and the curvature parameter, λ, which is taken as proportional to R, i.e. $\lambda = B\cdot R$, where B is an empirical parameter. Then, if n is an integer and r is the actual interatomic distance, the general form of the potential can be written as:

$$E(r) = \varepsilon(n+\lambda)/(n+\lambda-6)[-(R/r)^6 + 6/(n+\lambda)(R/r)^n \exp(\lambda(1-r/R))] \quad (1)$$

This general form takes two different forms in two approximations. If $n = 0$, the so-called 6-exp form is obtained:

$$E(r) = A\exp(-B\cdot r) - Cr^{-6} \quad (2)$$

$$A = 6\varepsilon/(\lambda-6); \quad B = \lambda/R; \quad C = \varepsilon\lambda R^6/(\lambda-6) \quad (3)$$

If $n = 12$ and $B = 0$, the so-called 6-12 form is obtained:

$$E = Fr^{-12} - Gr^{-6} \quad (4)$$

$$F = \varepsilon R^{12}; \quad G = 2\varepsilon R^6 \quad (5)$$

The two forms share a r^{-6} attractive term. The 6-exp form is a three-parameter curve, and the description of repulsion is more accurate, but computationally more expensive. The meaning of the parameter λ in the 6-exp form can be understood by writing ($E^* = E/\varepsilon$, $z = r/R$):

$$(\partial^2 E^*/\partial z^2)_{z=1} = 6\lambda(\lambda-7)/(\lambda-6) \quad (6)$$

Thus, λ controls the curvature of the potential at the minimum. For a 12-6 curve, $(\partial^2 E^* / \partial z^2)_{z=1} = 72$ (a constant); equating this value to the right-hand side of Eq. 6, it can be seen that a 12-6 curve corresponds to a 6-exp curve with a fixed value of $\lambda = 13.8$. Thus, 6-exp potentials whose λ parameter is close to this value can be recast in the 6-12 form without substantial loss of accuracy. A more detailed treatment is given in Chapter 4, Section 3.1.2.

7. COMPARING CALCULATED LATTICE ENERGIES WITH OBSERVED SUBLIMATION ENTHALPIES

The comparison of calculated lattice energies with observed sublimation enthalpies is burdened with certain difficulties. In the first place, the calculated internal energy of sublimation, $\Delta U_{\mathrm{c}}^{\mathrm{v}}$, ideally refers to 0 K, while heats of sublimation are measured at finite temperatures. Assuming as a reference $T = 300$ K, the correction to be applied is

$$\Delta U_{\mathrm{c}}^{\mathrm{v}}(300\ \mathrm{K}) = \Delta U_{\mathrm{c}}^{\mathrm{v}}(0\ \mathrm{K}) + \int_0^{300} \Delta C_{\mathrm{v}}\, \mathrm{d}T$$

The integral cannot be easily evaluated: while the ideal gas C_{v} is $3R$, independent of temperature, crystal heat capacities are variable from crystal to crystal and also depend on temperature, becoming very small at low T. A popular procedure is to assume instead the C_{v} of any crystal to be equal to $6R$ all the way from 0 to 300 – a questionable assumption, to say the least. Besides, any phase transition in the solid between 0 and 300 K (and there are often many) brings in an additional ΔU (phase).

Secondly, there is a difference between internal energies and enthalpies, as expressed by Eq. 7, page 71. This correction only applies at temperatures where one can plausibly assume $P = 1$ atm for the gas.

The combination of these two corrections leads to Eq. 8, page 71, yielding a $-2RT$ difference between calculated and experimental values.

On the other hand, sublimation enthalpies are usually determined at temperatures well above 300 K, where a significant vapor pressure can be measured for the solid. Reduction of these data to 300 K can be performed by Eq. 13, page 72, or as described by DeKruif in *J. Chem. Thermod.* **12**, 243 (1980), where ΔC_{P} is a temperature-dependent, adjusted parameter. This reduction, often disregarded by experimentalists, brings in a shift usually comparable to or larger than $2RT$: ΔC_{P}'s in reasonable temperature intervals are of the order of 50–80 J/(Kmol) for organic crystals (about 2 kcal/mole ΔC variation over 100 K). Sometimes, even the temperature of the X-ray structures on which the calculations (and the fit of potentials) are based may vary.

The contribution of zero-point energy to lattice energies can usually be neglected.

In conclusion, all the empirical corrections that can be proposed when matching calculations and experiment are rather unreliable, and, even more disturbingly, are well within the standard deviations of experimental sublimation enthalpies (see the references on solid-vapor equilibria in Section 5.10 of this chapter). A pragmatic attitude is to suppose that some of the theoretical difference between calculation and experiment is incorporated into the parameterization of the force field, and that an intrinsic significance threshold, when discussing agreement between calculation and experiment in this field, is about 2 kcal/mole, unless clear and independent thermodynamic evidence is given.

Chapter 4

Energy Minimization and Molecular Dynamics Calculations for Molecular Crystals

Bouke P. van Eijck, Loes M. J. Kroon-Batenburg and Jan Kroon
Department of Crystal and Structural Chemistry, Bijvoet Center for Biomolecular Research, Utrecht University

1. INTRODUCTION

In this chapter we shall assume that the potential energy of a system of atoms is known as a function of their coordinates:

$$U = U(\mathbf{r}_1, \mathbf{r}_2, \dots) \tag{1}$$

Then it should be possible to calculate many observable properties of the system. We here address this problem for crystals, with the emphasis on molecular crystals. In this particular case intramolecular forces are considerably larger than intermolecular ones, so it is meaningful to view the crystal as built from separate molecules.

The potential energy function U, often called the *force field*, exists if the Born–Oppenheimer approximation is valid. Then in principle one could calculate the force field *ab initio* by solving the electronic Schrödinger equation for fixed nuclear positions. Since the nuclear masses do not occur in that equation,

Theoretical Aspects and Computer Modeling Edited by A. Gavezzotti

the potential energy is invariant for isotopic substitution. In practice reliable *ab initio* calculations are only feasible for very simple crystals, although increasing computer power and program sophistication may change that situation in a few years time (see Chapter 2).

When no external fields are present it is also possible to write the potential energy as a function of the *interatomic distances* $l_{k,k'} \equiv |\mathbf{r}_k - \mathbf{r}_{k'}|$. Although the functional form is, of course, different from the definition in Eq. 1, we shall make no distinction in notation between the two formulations.

Section 2 in this chapter reviews how observable crystal properties can be derived from a force field in a general, unspecified form. In Section 3 we shall discuss the functional forms and the parametrization of force fields as they occur in the present literature. In the final section some recent applications will be discussed.

1.1 Crystals

For the convenience of the reader, and to introduce some notation, a short survey of the essentials of crystal structure is given. Detailed textbooks are plentiful, e.g. [1, 2].

In an ideal crystal, three vectors (**a**, **b**, **c**) define the unit cells of the periodic lattice. They are not necessarily orthogonal, so it will be convenient to express them in a Cartesian axes system where the components can be written as the matrix

$$\mathbf{C} \equiv \begin{pmatrix} a_x & b_x & c_x \\ a_y & b_y & c_y \\ a_z & b_z & c_z \end{pmatrix} \tag{2}$$

The inverse matrix is given by:

$$\mathbf{C}^{-1} = \begin{pmatrix} a_x^* & a_y^* & a_z^* \\ b_x^* & b_y^* & b_z^* \\ c_x^* & c_y^* & c_z^* \end{pmatrix} \equiv \mathbf{C}^* \tag{3}$$

where $\mathbf{a}^*$, $\mathbf{b}^*$, $\mathbf{c}^*$ are the *reciprocal axes*. They satisfy relations like $\mathbf{a}^* = [\mathbf{b} \times \mathbf{c}]/V$ where the denominator is the cell volume $V = \mathbf{a}\cdot[\mathbf{b} \times \mathbf{c}] = \det \mathbf{C}$. The orientation of the Cartesian axes system can be freely chosen, for instance

$$a_y = a_z = b_z = 0 \tag{4}$$

which makes $\mathbf{C}$ as well as $\mathbf{C}^*$ upper triangular. Six parameters are left to define the unit cell. Usually these are transformed into the cell edges a, b, c and the cell angles α, β, γ. Alternatively, the six independent elements of the *metric matrix* $\mathbf{G} = \mathbf{C}^{\mathrm{T}}\mathbf{C}$ can be used.

Each unit cell can be numbered with three integers l_1, l_2, l_3 (l for short). The equilibrium position of the kth atom in that cell can be expressed as:

$$r^{\mathrm{e}}_{lk} = l_1\mathbf{a} + l_2\mathbf{b} + l_3\mathbf{c} + \mathbf{r}^{\mathrm{e}}_k \tag{5}$$

where the translational symmetry ensures that $\mathbf{r}^{\mathrm{e}}_k$, the equilibrium position with respect to the cell origin, is independent of l. However, due to thermal motion the actual atomic positions $\mathbf{r}_{lk}$ can be different in each cell:

$$\mathbf{r}_{lk} = \mathbf{r}^{\mathrm{e}}_{lk} + \Delta\mathbf{r}_{lk} \tag{6}$$

Equilibrium positions are defined by the condition that forces on any atom are zero:

$$F^{\mathrm{e}}_{lk\alpha} = -\left(\frac{\partial U}{\partial r_{lk\alpha}}\right)_{\mathrm{e}} = 0 \tag{7}$$

(In this text, α, β, ... will be used as indices to denote one of the x, y, z components along the coordinate axes in a Cartesian axes system; no confusion with the cell angles should occur.) The equilibrium coordinates that satisfy this equation depend on the unit cell parameters $\mathbf{C}$, which in their turn are determined by the boundary conditions (temperature and pressure) imposed externally on the crystal.

It is customary in crystallography to express vectors $\mathbf{r}$ in *fractional coordinates* denoted by x, y, z:

$$\mathbf{r} = x\mathbf{a} + y\mathbf{b} + z\mathbf{c} \tag{8}$$

This choice of notation is unfortunate since it suggests strongly that x, y, z are expressed in a Cartesian axes system, which is not the case. We shall prefer to denote these fractional coordinates by a vector $\mathbf{s}$; the relation with the Cartesian coordinates is given by $\mathbf{r} = \mathbf{Cs}$ or, in components:

$$r_\alpha = \sum_g C_{\alpha g} s_g \tag{9}$$

where the index g runs over a, b, c. The energy can then be expressed as a function of $\mathbf{s}_{lk}$ and $\mathbf{C}$.

Within each unit cell additional symmetry may exist. There are 230 possible combinations of symmetry elements, the *space groups*. For instance, in the most frequently occurring space group for organic molecules ($P2_1/c$) the cell angles α and γ are restrained to 90°, and each atom with equilibrium fractional coordinates (x, y, z) is accompanied by three identical atoms at positions $(-x, -y, -z)$, $(-x, y + \frac{1}{2}, -z + \frac{1}{2})$, $(x, -y + \frac{1}{2}, z + \frac{1}{2})$. These so-called *equivalent positions*, together with a wealth of other detailed information, can be found in the *International Tables for X-ray Crystallography* [3] for each space group. More formally, the ith symmetry operation transforms a vector $\mathbf{s}$ in the *asymmetric unit* into an equivalent position $\mathbf{s}_i$ following:

$$\mathbf{s}_i = \mathbf{M}_i\mathbf{s} + \mathbf{t}_i \tag{10}$$

where $\mathbf{M}_i$ is a 3×3 matrix with elements 0, 1 or -1 and $\mathbf{t}_i$ a translation vector with elements that are integers or fractions. In Cartesian coordinates this formula can be written as:

$$\mathbf{r}_i = \mathbf{C}\mathbf{M}_i\mathbf{C}^{-1}\mathbf{r} + \mathbf{C}\mathbf{t}_i \equiv \mathbf{M}_i'\mathbf{r} + \mathbf{C}\mathbf{t}_i \tag{11}$$

It can be seen that $\mathbf{M}_i'$ represents a pure rotation and/or inversion in Cartesian axes, so it must be independent of $\mathbf{C}$. For triclinic, monoclinic and the three orthogonal systems we even have simply $\mathbf{M}_i' = \mathbf{M}_i$. This can be shown from the equivalent relation $\mathbf{C}\mathbf{M}_i = \mathbf{M}_i\mathbf{C}$, where at least one of the matrices $\mathbf{M}_i$ or $\mathbf{C}$ is diagonal, by considering all possible situations. However, although Pertsin and Kitaigorodsky [4, p. 165] say this relation is obvious, it is not valid for trigonal and hexagonal space groups.

2. CALCULATION OF OBSERVABLE PROPERTIES

2.1 Background

Detailed structural information on the cell dimensions and the positions of the atoms in the unit cell can be obtained by X-ray or neutron diffraction [1, 2]. Repeating the experiment at different temperatures gives the thermal expansion coefficients. Diffraction analysis also produces the mean square amplitudes of vibration $\langle \Delta r_{k\alpha} \Delta r_{k\beta} \rangle$ for each atom k. These amplitudes are related to the possible vibrational modes in the crystal, intramolecular as well as intermolecular, which can be studied by spectroscopic methods and which are also responsible for thermal diffuse scattering of X-rays [5].

Calorimetric measurements can give the specific heats and the enthalpies of sublimation which are directly related to thermodynamic quantities. Measurement of the elasticity tensor, preferably as a function of the temperature, also provides important data; for a review see Kitaigorodsky [6]. Electric effects (dielectric constants, piezoelectricity, etc.) have been treated in detail by Born and Huang [7].

Geometric details of over 160 000 organic and metal–organic crystal structures can be retrieved from the computerized Cambridge Structural Database [8]. Unfortunately, atomic displacement parameters cannot be found so easily. Enthalpies of sublimation have been collected by Chickos [9] (see also Chapter 3). Elastic, piezoelectric and related constants of crystals are discussed and collected in Volume III-11 of the Landolt–Börnstein series [10] (with an update in Volume III-18).

If a force field is able to reproduce these and other observations quantitatively for a variety of substances, one may have some confidence in predictions for other substances, or even other properties, where no experimental data are available. The classical road from theory to observable properties is by analytical development. However, the mathematical equations tend to become quite complex, even for simple harmonic force fields. Computer simulation can eliminate some of the difficulty; in many cases even anharmonicity hardly poses any special problems there.

In this section we shall give an introduction to energy-minimization and molecular-dynamics calculations without putting any limitations on the form of the force field. Although the principles of these methods are quite simple, a good understanding of the underlying theory remains desirable. So we shall first provide some background on statistical thermodynamics and lattice vibrations; however, the casual reader might skip this part and take the theory for granted.

2.1.1 *Statistical Thermodynamics*

The calculation of observable properties from the potential energy function is the subject of statistical thermodynamics [11]. For a given temperature T and volume V the free energy A can be found from

$$A = -kT \ln Q \tag{12}$$

Here the *partition function* Q is defined by

$$Q = \sum_j e^{-E_j/kT} \tag{13}$$

where j denotes a quantum state with energy E_j. This is the fundamental equation that connects the macroscopic quantity A with a sum over all accessible microscopic states of the system.

In molecular dynamics, the subject of Section 2.3, it is assumed that the system can be described classically. The energy E of N particles is then expressed as a function of any set of general coordinates q_i and their time derivatives $\dot{q}_i$. The generalized momentum is defined as:

$$p_i = \frac{\partial E(\mathbf{q}, \dot{\mathbf{q}})}{\partial \dot{q}_i} \tag{14}$$

The Hamiltonian, $H(\mathbf{p}, \mathbf{q})$, is just the energy expressed in the $3N$ coordinates $\mathbf{q}$ and the $3N$ momenta $\mathbf{p}$, the *phase space*. In cartesian coordinates the q_i correspond to $\mathbf{r}_k$ and the p_i to $m_k \mathbf{v}_k$ ($i = 1, \ldots, 3N$, $k = 1, \ldots, N$).

The partition function is then given by:

$$Q = \frac{\int\int e^{-H(\mathbf{p},\mathbf{q})/kT}\, d\mathbf{p}\, d\mathbf{q}}{h^{3N} N_A! N_B! \ldots} \tag{15}$$

where h is Planck's constant, N_A denotes the number of identical atoms of type A, etc. For crystals the denominator is simply h^{3N} if each coordinate is restrained to a small region around one equilibrium position, so that no permutations of identical atoms are allowed. Since the Hamiltonian is a quadratic form of the momenta, the integral over **p** can always be singled out and evaluated analytically; in Cartesian coordinates one gets:

$$Q = \prod_{k=1}^{N} \left(\frac{2\pi m_k kT}{h^2} \right)^{3/2} \int \ldots \int e^{-U/kT} \, d\mathbf{r}_1 \ldots d\mathbf{r}_N \tag{16}$$

where k, the Boltzmann constant, should not be confused with the atomic index k.

The *phase-space probability density* is given by:

$$\pi(\mathbf{p}, \mathbf{q}) = \frac{e^{-H(\mathbf{p},\mathbf{q})/kT}}{\displaystyle\iint e^{-H(\mathbf{p},\mathbf{q})/kT} \, d\mathbf{p} \, d\mathbf{q}} \tag{17}$$

from which the average value of an observable X which depends on **p** and **q** follows:

$$\langle X \rangle = \iint X(\mathbf{p}, \mathbf{q}) \pi(\mathbf{p}, \mathbf{q}) \, d\mathbf{p} \, d\mathbf{q} = \frac{\displaystyle\iint X(\mathbf{p}, \mathbf{q}) \, e^{-H(\mathbf{p},\mathbf{q})/kT} \, d\mathbf{p} \, d\mathbf{q}}{\displaystyle\iint e^{-H(\mathbf{p},\mathbf{q})/kT} \, d\mathbf{p} \, d\mathbf{q}} \tag{18}$$

For example, the kinetic energy is found to be $\frac{3}{2}NkT$, irrespective of the form of the potential energy. For the entropy S in a non-diffusive system like a crystal one can derive the expression:

$$S = -k \iint \pi(\mathbf{p}, \mathbf{q}) \ln [h^{3N} \pi(\mathbf{p}, \mathbf{q})] \, d\mathbf{p} \, d\mathbf{q} \tag{19}$$

For a crystal at constant volume there are many configurations with low energy, each characterized by different cell parameters and equilibrium structures, that might conceivably contribute to the integral in Eq. 16. In any reasonable time scale, however, there will be no interconversion between such structures unless they are very similar: in that case the crystal is disordered. Usually differences are too large and energy barriers too high, giving rise to the possibility of several discrete polymorphs. For each one we must calculate the free energy separately, retaining in the partition function only the accessible states for that polymorph. The shape of the unit cells can be modified to a certain extent by external conditions. It follows that the free energy becomes a function of T and **C** rather than T and V.

The equations given so far are exact for the entire system. But a macroscopic system is much too large to handle and, of course, the translational symmetry of the crystal is used. However, such calculations are not without pitfalls. Consider, for instance, a one-dimensional chain of $N+1$ atoms linked by N harmonic springs around an equilibrium length l^{e}:

$$U = \tfrac{1}{2} f \sum_{k=1}^{N} (x_{k+1} - x_k - l^{\mathrm{e}})^2 \tag{20}$$

The force on atom k in an inner section of the chain is given by:

$$F_k = -\frac{\partial U}{\partial x_k} = f(x_{k+1} - x_k - l^{\mathrm{e}}) - f(x_k - x_{k-1} - l^{\mathrm{e}}) \tag{21}$$

There is obviously equilibrium, $F_k = 0$, if the interatomic distances are equal to the cell dimension a, e.g. $x_k^{\mathrm{e}} = ka$. The total energy is then $\frac{1}{2} f N(a - l^{\mathrm{e}})^2$, and:

$$\frac{\partial U}{\partial a} = fN(a - l^{\mathrm{e}}) \tag{22}$$

However, if one were to calculate this derivative by the chain rule:

$$\frac{\partial U}{\partial a} = \sum_{k=1}^{N+1} \frac{\partial U}{\partial x_k} \frac{\partial x_k}{\partial a} \tag{23}$$

an erroneous result of zero would be found since the forces F_k are zero in equilibrium. The explanation of the paradox is that the end atoms $k = 1$ and $k = N + 1$, where the forces are not zero, were not taken into account. Here the forces should be compensated by the external pressure on the crystal. The difficulty disappears if the energy is defined as a function not of coordinates, but of interatomic distances $l_k \equiv x_{k+1} - x_k$:

$$U = \tfrac{1}{2} f \sum_{k=1}^{N} (l_k - l^{\mathrm{e}})^2 \tag{24}$$

Now the chain rule

$$\frac{\partial U}{\partial a} = \sum_{k=1}^{N} \frac{\partial U}{\partial l_k} \frac{\partial l_k}{\partial a} \tag{25}$$

gives the correct result, Eq. 22.

The pressure of a crystal is not a scalar quantity, but a tensor **P** defined as follows [11, p. 382]. A surface, defined by an outward vector **N** that is perpendicular to that surface and with absolute value equal to the surface area, experiences a force **F** with components:

$$F_\alpha = \sum_\beta P_{\alpha\beta} N_\beta \tag{26}$$

Derivations of expressions for the pressure are not readily found in the literature, so they are given in an appendix. The results are as follows. The pressure components can be found from the free energy by:

$$P_{\alpha\beta} = -\frac{1}{V}\sum_{g}\frac{\partial A}{\partial C_{\alpha g}}C_{\beta g} \tag{27}$$

This is a generalization of the well-known equation $P = -\partial A/\partial V$ for the isotropic case. It requires knowledge of the free energy as a function of T and $\mathbf{C}$, which is usually not easily available.

For molecular simulations, where classical mechanics is assumed, it is more practical to express the pressure in the coordinates $\mathbf{r}_k$ and the velocities $\mathbf{v}_k$:

$$P_{\alpha\beta}V = \sum_{k} m_k\langle v_{k\alpha}v_{k\beta}\rangle + W_{\alpha\beta} \tag{28}$$

where $\mathbf{W}$ is the *virial*:

$$W_{\alpha\beta} = -\sum_{k}\left\langle\frac{\partial U}{\partial r_{k\alpha}}r_{k\beta}\right\rangle \tag{29}$$

or, equivalently,

$$W_{\alpha\beta} = -\sum_{g}\left\langle\frac{\partial U(\mathbf{s},\mathbf{C})}{\partial C_{\alpha g}}\right\rangle C_{\beta g} \tag{30}$$

where U must be expressed as a function of $\mathbf{C}$ and the *fractional* coordinates of the atoms.

The brackets in these equations denote the average over the possible configurations of the system, as obtained from molecular dynamics (Section 2.3) or Monte Carlo (Section 2.4) simulations, and corresponding to the ensemble average in Eq. 18. It is possible to replace the first term in Eq. 28 by $NkT\delta_{\alpha\beta}$, where N denotes the number of atoms and $\delta_{\alpha\beta}$ is Kronecker delta ($\delta_{\alpha\beta} = 0$ for $\alpha \neq \beta$, $\delta_{\alpha\alpha} = 1$). Note that some authors use different definitions for the virial.

For reasons analogous to those given above in our cautionary discussion of a one-dimensional atom chain, Eq. 29 cannot be applied if the atoms considered are only the repeating unit of a periodic system. For a detailed discussion of the isotropic case, see Ref. [12]. It is necessary to express energy and virial in the interatomic distances $l_{k,k'}$:

$$W_{\alpha\beta} = -\sum_{k}\sum_{k'>k}\left\langle\frac{1}{l_{k,k'}}\frac{\partial U}{\partial l_{k,k'}}(r_{k\alpha} - r_{k'\alpha})(r_{k\beta} - r_{k'\beta})\right\rangle \tag{31}$$

This equation shows that the pressure tensor is symmetric: $P_{\alpha\beta} = P_{\beta\alpha}$, a property that can also be found from macroscopic considerations [13, p. 115].

It is possible to calculate the pressure on the basis of M molecules rather than N atoms. The relevant equation is:

$$P_{\alpha\beta}V = MkT\delta_{\alpha\beta} - \sum_{k}\sum_{k'>k}\left\langle \frac{1}{l_{k,k'}}\frac{\partial U}{\partial l_{k,k'}}(R_{t\alpha} - R_{t'\alpha})(r_{k\beta} - r_{k'\beta})\right\rangle \tag{32}$$

where k and k' belong to different molecules t and t', respectively, and $\mathbf{R}_t$ refers to any positional vector in molecule t, and is usually taken to be the position of the molecular centre of mass.

If (as usual) the crystal is in an environment with isotropic pressure P^0, the cell parameters $\mathbf{C}$ will adjust themselves in such a way as to make $P_{\alpha\beta} = P^0\delta_{\alpha\beta}$. With the axes choice of Eq. 4 the equations to be solved are simply:

$$\frac{\partial A}{\partial C_{\alpha\alpha}} = -\frac{P^0 V}{C_{\alpha\alpha}}$$
$$\frac{\partial A}{\partial C_{\alpha g}} = 0 \quad (\alpha < g) \tag{33}$$

In the calculations, the difference between atmospheric and zero pressure can often be neglected with good approximation. When the stability of different polymorphs is compared at constant pressure, the criterion should be the difference between the Gibbs energies $G = A + PV$. Again, the PV term is usually negligible.

2.1.2 Lattice Vibrations

The temperature dependence of crystal properties arises through atomic vibrations. To find the free energy one has to solve the Schrödinger equation for the displacements of the atoms from their equilibrium positions ($\Delta\mathbf{r}_{lk}$ in Eq. 6). Since these displacements are of small amplitude, it is useful to expand the potential energy into a power series:

$$U = U^{\mathrm{e}} + \sum_{lk\alpha}\left(\frac{\partial U}{\partial r_{lk\alpha}}\right)_{\mathrm{e}}\Delta r_{lk\alpha} + \tfrac{1}{2}\sum_{lk\alpha}\sum_{l'k'\beta}\Delta r_{lk\alpha}\Phi_{lk\alpha,l'k'\beta}\Delta r_{l'k'\beta} \tag{34}$$

where the force constants Φ are defined by:

$$\Phi_{lk\alpha,l'k'\beta} = \left(\frac{\partial^2 U}{\partial r_{lk\alpha}\partial r_{l'k'\beta}}\right)_{\mathrm{e}} \tag{35}$$

The linear term in $\Delta\mathbf{r}$ is zero by virtue of Eq. 7, and higher-order terms have been neglected, i.e. the vibrations are assumed to be harmonic. The problem is like the determination of the normal vibrations of a molecule, except that the number of atoms is so large that a straightforward approach is entirely infeasible.

Fortunately the translational symmetry of the crystal can be used to simplify the problem [5, 7, 14], with the following results.

Consider a small part of a crystal, which is a parallelepiped of unit cells with periodic boundary conditions, and define a *wave vector* $\mathbf{q}$:

$$\mathbf{q} = 2\pi\left(\frac{l_1}{N_1}\mathbf{a}^* + \frac{l_2}{N_2}\mathbf{b}^* + \frac{l_3}{N_3}\mathbf{c}^*\right) \tag{36}$$

where N_1 is the number of unit cells in the $\mathbf{a}$ direction and l_1 is an integer that can take values from 1 to N_1; likewise for the directions $\mathbf{b}$ and $\mathbf{c}$. Then it can be shown that the normal vibrations for each value of $\mathbf{q}$ can be treated separately, by finding the eigenvectors of the *dynamical matrix*:

$$D_{k\alpha,k'\beta}(\mathbf{q}) = \frac{1}{\sqrt{m_k m_{k'}}}\sum_{l'} \Phi_{lk\alpha,l'k'\beta} \exp\left[\mathrm{i}\mathbf{q}\cdot(\mathbf{r}^{\mathrm{e}}_{l'k'} - \mathbf{r}^{\mathrm{e}}_{lk})\right] \tag{37}$$

Since the crystal has translational symmetry and the summation is over all unit cells considered, the right-hand side of this equation does not depend on l. If there are N_{cell} atoms in the unit cell, the dimension of $\mathbf{D}$ will be $3N_{\mathrm{cell}} \times 3N_{\mathrm{cell}}$. The eigenvalue problem can be formulated as:

$$\sum_{k'\beta} D_{k\alpha,k'\beta}(\mathbf{q}) e_{k'\beta,j}(\mathbf{q}) = 4\pi^2 \nu_j^2(\mathbf{q}) e_{k\alpha,j}(\mathbf{q}) \tag{38}$$

where the normalization condition for the eigenvectors is:

$$\sum_{k\alpha} e^*_{k\alpha,j}(\mathbf{q}) e_{k\alpha,j'}(\mathbf{q}) = \delta_{jj'} \tag{39}$$

By solving these equations the harmonic frequencies $\nu_j (j = 1, \ldots, 3N_{\mathrm{cell}})$ for the relevant value of $\mathbf{q}$ are found. The corresponding energy levels of the lattice vibrational modes are $h\nu_j(v_j + \frac{1}{2})$ where $v_j = 0, 1, 2, \ldots$. Then, it is easy to obtain an expression for the free energy [11, p. 197]:

$$A = U^{\mathrm{e}} + \sum_{\mathbf{q}}\sum_{j}\left[\tfrac{1}{2}h\nu_j(\mathbf{q}) + kT \ln\left(1 - \mathrm{e}^{-h\nu_j(\mathbf{q})/kT}\right)\right] \tag{40}$$

In a molecular crystal containing nonlinear molecules one finds for each molecule in the unit cell six intermolecular vibrational modes, with low frequencies (typically below 200 cm^{-1}) that depend strongly on $\mathbf{q}$. Three of these can be regarded as translational modes (their frequencies will approach zero for $\mathbf{q} \to \mathbf{0}$) and three as librational modes of rigid molecules. (For linear molecules there will be only two librational modes, and for an atomic crystal none.) The other modes, which have much higher frequencies, will resemble the intramolecular vibrations in a free molecule, and their frequency dependence on $\mathbf{q}$ will be small.

When N_1, N_2 and N_3 become so large that $\mathbf{q}$ can be regarded as a continuous variable, the results are independent of these numbers, i.e. of the shape of the crystal, as should be the case. It is, of course, impossible to do the calculations for every value of $\mathbf{q}$, and in practice a limited set is taken that is homogeneously distributed within one unit cell of the reciprocal lattice (the *first Brillouin zone*). When results depend only on the frequencies (like energy and free energy), the summation over $\mathbf{q}$ and j can be replaced by an integration over a frequency distribution $g(\nu)$. This should be normalized according to

$$\int g(\nu)\,\mathrm{d}\nu = 3N_{\text{cell}}N_1N_2N_3 = 3N \tag{41}$$

where N is the total number of atoms in the system. Approximate expressions for the intermolecular contributions to $g(\nu)$ are the well-known formulae proposed by Einstein or by Debye [11].

From X-ray measurements the vibrational amplitudes $\langle \Delta r_{k\alpha}\Delta r_{k\beta}\rangle$ can be obtained. The results can be compared with the theoretical expression [5]:

$$\langle \Delta r_{k\alpha}\Delta r_{k\beta}\rangle = \sum_{\mathbf{q}}\sum_{j}\frac{h}{4\pi^2\nu_j(\mathbf{q})m_kN}|e^*_{k\alpha,j}(\mathbf{q})e_{k\beta,j}(\mathbf{q})|\{\tfrac{1}{2}+(\mathrm{e}^{h\nu_j(\mathbf{q})/kT}-1)^{-1}\} \tag{42}$$

Thus detailed knowledge of $\mathbf{e}_{kj}$ is necessary to calculate these quantities, but from Eq. 39 it follows that the sum $\sum_k m_k\langle \Delta\mathbf{r}_k^2\rangle$ can be found even when only the frequency distribution $g(\nu)$ is known.

In Eq. 40 the dependence of the free energy on temperature and atomic masses is only introduced through the second term on the right-hand side. If the crystal vibrations were purely harmonic, U would have no third derivatives to the coordinates and so the force constants $\mathbf{\Phi}$ and the frequencies ν would be independent of the coordinates and, therefore, of the cell parameters. So in Eq. 27 the free energy A could be replaced by the equilibrium energy U^{e}, and the pressure would be independent of temperature; the same would hold for the crystal structure. In a classical treatment it is obvious from Eqs 16 and 27 that pressure is independent of the atomic masses; any change of the crystal structure upon isotopic substitution must be caused by quantum effects and cannot be found by classical simulations.

The theory discussed above is also useful for the interpretation of inelastic neutron scattering and thermal diffuse X-ray scattering. However, it is incomplete since anharmonic effects are neglected and, in any case, such a crystal would have no thermal expansion [5, 13, 14]. It is clear that a more complete theory would become extremely complex. Approximate solutions are usually based on the Debye temperature Θ_{D} which is h/k times the cutoff frequency for a parabolic frequency-distribution curve [11]. This quantity is considered as a function of $\mathbf{C}$ and T, to be determined experimentally. For a detailed development of this *quasi-harmonic* approximation see Ref. [6]. In an alternative approach, the *cell model*, each crystal cell is assumed to be independent of the others. The two approximations should work best for low and high temperatures, respectively [4].

2.2 Energy Minimization: Molecular Mechanics

In order to find the crystal structure parameters from the force field, free energy derivatives with respect to cell parameters are needed (Eq. 33). As it turns out, cell parameters are not very sensitive to temperature and pressure. So one possibility will be to neglect the effect of lattice vibrations, including the zero-point energy. Then the free energy is equal to the equilibrium potential energy U^{e}, which is a function of $\mathbf{C}$ through all the equilibrium coordinates $\mathbf{r}^{\mathrm{e}}_{lk}$ in the system. In practice, of course, only one asymmetric unit is considered and U^{e} can be expressed as a function of unit cell parameters and of equilibrium coordinates in the asymmetric unit:

$$U^{\mathrm{e}} = U^{\mathrm{e}}(\mathbf{C}, \mathbf{r}^{\mathrm{e}}_1, \mathbf{r}^{\mathrm{e}}_2, \ldots) \tag{43}$$

The two conditions, Eq. 33 (with P^0 approximated by zero) and Eq. 7, now combine to the simple criterion:

$$\frac{\partial U^{\mathrm{e}}}{\partial p_i} = 0 \tag{44}$$

where p_i is the set of parameters $C_{\alpha g}$, $r^{\mathrm{e}}_{k\alpha}$. Performing this energy minimization, also called *molecular mechanics*, is the subject of this section.

The exact number of parameters depends on space-group symmetry. As stated in Section 1.1, one can always set $a_y = a_z = b_z = 0$. In the orthorhombic system, for example, the cell is orthogonal ($\alpha = \beta = \gamma = 90°$), which leads to a diagonal $\mathbf{C}$ matrix. In the most simple space group (P1) there is only translational symmetry, so there is no natural choice of origin and one of the atoms can be placed at will. In space group *P*2, where the *b*-axis becomes a twofold rotation axis, only the *y*-coordinate of one atom can be chosen freely, and there are the additional conditions $\alpha = \gamma = 90°$ (or $b_x = c_y = 0$) for a monoclinic space group (see Chapter 1).

The number of mathematical solutions of Eq. 44, corresponding physically to different polymorphs, is enormous. Finding the one with lowest energy is a great challenge (see Section 4.3 and Chapter 6). In this section we shall assume that an initial guess for the parameter values, p^0_i, is available and that the set p^{e}_i which corresponds to the energy minimum is sought. One application is to start from the experimental crystal structure; ideally the parameters should satisfy Eq. 44, but in practice a shift will occur which is a measure of the inadequacy of the force field used.

To summarize, a local minimum of the function $U^{\mathrm{e}}(p_i)$, rather than the global minimum, is sought. Even this restricted problem is not simple. It is a very general one, discussed in a vast amount of literature [15, 16, 17]. Therefore only a short review will be given.

2.2.1 First-Derivative Methods

Let us suppose that only the first derivatives of the energy parameters are available:

$$g_i \equiv \frac{\partial U^e}{\partial p_i} \tag{45}$$

Minimization is usually an iterative process along successive search directions $\mathbf{z}$ in parameter space:

$$\mathbf{p} = \mathbf{p}^0 + c\mathbf{z} \tag{46}$$

In that direction U^e is a function of c only, and it is possible to perform a *line minimization* with respect to c:

$$\frac{dU^e}{dc} = \sum_i g_i z_i = 0 \tag{47}$$

In the *steepest descent* method the search direction is equal to minus the gradient of the energy at the starting point: $\mathbf{z} = -\mathbf{g}$. In the simplest version of this method the line minimization is not carried out, and c is just a positive constant to be judiciously chosen according to the nature of the problem. Obviously there is no reason why the energy minimum should be reached in one step, so the process has to be repeated many times. When c is small enough the energy must decrease, and the method is slow but safe. It will always converge to the nearest local minimum. Alternatively, the line minimization (Eq. 47) is performed in each cycle. Then the number of iterations decreases, but each one is more time-consuming.

The number of iterations can be decreased even further by using the *conjugate gradients* method. Here line minimizations are essential. In the beginning the search direction is again taken equal to minus the gradient, but in each following new cycle it is updated from the previous one as follows:

$$\mathbf{z}^{\text{new}} = -\mathbf{g}^{\text{new}} + \mathbf{z}^{\text{previous}} \frac{\sum_i (g_i^{\text{new}})^2}{\sum_i (g_i^{\text{previous}})^2} \tag{48}$$

Some variations of this equation are possible [15, 16]. When the energy is a quadratic function of the parameters, it can be shown that this process converges in just as many cycles as there are parameters to be optimized.

In all methods it may be possible to increase the performance by proper choice (scaling or preconditioning) of the parameters. This is obvious, since their effects on the energy can vary over orders of magnitude; the parameters could even have different physical dimensions.

2.2.2 Second-Derivative Methods

The second derivatives of the energy to the parameters are written as:

$$F_{ij} = \frac{\partial^2 U^{\mathrm{e}}}{\partial p_i \partial p_j} \tag{49}$$

This resembles Eq. 35 which is needed for the calculation of lattice vibrations, but there all atom positions were allowed to change individually, whereas in Eq. 49 the crystal symmetry is usually enforced. If the second derivatives are available, the parameter shifts can be elegantly expressed by the Newton–Raphson equation:

$$\mathbf{p} = \mathbf{p}^0 - \mathbf{F}^{-1}\mathbf{g} \tag{50}$$

For a quadratic function the results should be exact in one iteration, but the calculation of **F** can be quite tedious, and the solution of Eq. 50 also takes some time. A further disadvantage of this method is that it may converge to a maximum or a saddle point rather than a minimum if **F** is not positive definite. Therefore, it is advisable to perform a few steepest-descent cycles first. For refinements of the method (modified, discrete, quasi-, truncated Newton) see [16]; some of these eliminate the need for explicit calculation of the second derivatives.

2.3 Classical Simulation: Molecular Dynamics

The scope of energy-minimization methods, as described in the previous section, is somewhat limited. For one thing, the temperature plays no role. At the expense of greater effort, especially in computer cost, more information can be obtained from *molecular dynamics*. Here the force field is used to construct the classical trajectories of the atoms as they move around, and properties that depend on temperature and on time can be obtained. The method will be discussed here for crystals, but it is more widely used to simulate liquids; in fact, for that state of matter it is the only way to obtain macroscopic properties from a microscopic model. Extensive reviews are available [18, 19] (see Chapter 5 for applications to small clusters, and Chapter 6 for application to crystal-structure prediction).

Assuming that classical mechanics can be applied, Newton's equations of motion are set up for each atom k:

$$m_k \mathbf{a}_k = m_k \frac{\mathrm{d}^2 \mathbf{r}_k}{\mathrm{d}t^2} = \mathbf{F}_k = -\nabla_k U(\mathbf{r}_1, \mathbf{r}_2, \ldots) \tag{51}$$

The most simple, yet surprisingly adequate, way to solve these equations numerically is the so-called *leap-frog* algorithm: given the coordinates $\mathbf{r}_k$ at time t and the velocities $\mathbf{v}_k$ at time $t - \frac{1}{2}\Delta t$, the equations are:

$$\begin{aligned} \mathbf{v}_k(t + \tfrac{1}{2}\Delta t) &= \mathbf{v}_k(t - \tfrac{1}{2}\Delta t) + \mathbf{a}_k(t)\Delta t \\ \mathbf{r}_k(t + \Delta t) &= \mathbf{r}_k(t) + \mathbf{v}_k(t + \tfrac{1}{2}\Delta t)\Delta t \end{aligned} \tag{52}$$

Here Δt is a time step, small enough to keep velocities and accelerations fairly constant between t and $t + \Delta t$; the order of Δt is 10^{-15} s. The result is a time history of the system: $\mathbf{r}_k(t)$ and $\mathbf{v}_k(t)$ for an interval of perhaps 1 ns (one million iterations of Eqs 52).

On today's workstations such a calculation is feasible for systems up to at least 1000 atoms. To avoid surface effects, periodic boundaries are introduced. For a liquid this is a surprisingly good approximation, and for a crystal it comes quite naturally. The system is taken to be a supercell of the crystal: a parallelepiped of a limited number of unit cells, usually chosen in such a way to obtain a *computational box* with comparable lengths along each axis.

The assumption that classical mechanics apply is quite questionable. Intramolecular vibrations have frequencies up to 4000 cm^{-1}, which is about twenty times kT at room temperature. Quantum effects become noticeable when the spacings between energy levels are comparable to or larger than kT, so classical mechanics would seem to be inappropriate. One way to reduce the problem is to consider the molecules as rigid units in the crystal, so as to retain only the intermolecular vibrations that have low frequencies. For simple prototypes of molecular crystals this is a very good assumption, but for more complex substances essential degrees of freedom may be lost. A compromise can be to consider only bond lengths as fixed, but to allow freedom in bond angles and dihedral angles. Technically this can be done by using a reduced set of coordinates $\mathbf{q}$ at the expense of complications in the determination of expectation values (see Ref. [19] for details but note that the right-hand side of Eq. (2.155) should be inverted and that m in Eqs (2.156) and (2.157) should be replaced by m^{-1}). An easier method is to use the so-called SHAKE algorithm [20, 21] by which atomic displacements are calculated under the constraint of constant bond lengths. Even without such constraints it is usually assumed that the classical trajectories give a realistic description of the system, although quantum corrections [19, 22] may be important for some applications.

The resulting trajectory $\mathbf{r}_k(t)$, $\mathbf{v}_k(t)$ should be a good basis to obtain observable quantities by simple averaging. For instance, after n time steps with N atoms the temperature follows from

$$3NkT = \left\langle \sum_{k=1}^{N} m_k \mathbf{v}_k^2 \right\rangle = \sum_{t=1}^{n} \sum_{k=1}^{N} m_k \mathbf{v}_k^2(t)/n \tag{53}$$

To obtain $\mathbf{v}(t)$ the average between $\mathbf{v}(t - \frac{1}{2}\Delta t)$ and $\mathbf{v}(t + \frac{1}{2}\Delta t)$ is taken, for $\mathbf{v}^2(t)$

the average of the squares should be used [23] rather than the square of the average, which is not always implemented correctly. For a molecular system it is possible to use Eq. 53 with different meanings of the symbols: the summation is then taken over N molecules with molecular mass m_k and centre-of-mass velocity $\mathbf{v}_k$.

In Eq. 53, $3N$ is the number of degrees of freedom of the system. However, motion of the system as a whole should not contribute to its temperature, so the velocities have to be corrected in such a way that the centre of gravity does not move. In this way three degrees of freedom are lost, and in Eq. 53 it is necessary to replace $3N$ by $3N - 3$. This number has to be further diminished by one for every bond length that is constrained.

The pressure is found likewise from Eqs 28 and 31, or from Eq. 32. Of course, these and other averages should be independent of the arbitrarily chosen initial values for coordinates and velocities. It is advisable to discard as *equilibrium time* the first part of the trajectory, say 10 ps.

The reliability of the results is very difficult to assess. There are systematic errors caused by inadequacies of the force field, the neglect of quantum effects, and the limited number of atoms in the system. There are also random errors due to incomplete sampling in a finite number of time steps. Here the usual methods to determine root-mean-square errors from a set of n data have to be used with caution, since subsequent points on the trajectory are highly correlated. One solution is to study the autocorrelation of the data [24]. An easier method is to make block averages over, say, 1000 time steps and to consider such averages as independent data [19, 25]. A nonrandom error due to inadequate simulation time can be that slow processes (on the 10^{-9} s time scale, like conformational transitions or other crystal disorder) are not encountered or are inadequately sampled.

2.3.1 *Simulations at Constant Temperature and Pressure*

The molecular dynamics simulations described above imply constant energy and volume, from which temperature and pressure follow. An essential test for a computer program is, therefore, that the energy:

$$E = \tfrac{1}{2}\sum_{k=1}^{N} m_k \mathbf{v}_k^2 + U(\mathbf{r}_1, \mathbf{r}_2, \ldots, \mathbf{r}_N) \tag{54}$$

must be constant at every time step. Likewise, linear momentum should be conserved. In practice these conditions will not be fulfilled exactly due to round-off errors, the use of a finite time step, and in some cases the use of a discontinuous cutoff radius for the interatomic interactions (see Section 3.3). The energy may show short-time fluctuations as well as a slow long-term drift.

For this reason, but also because it is physically more attractive, one would prefer to be able to select the parameters temperature (T^0) and pressure (P^0) rather than energy and volume. To this end other statistical-mechanical ensembles can be constructed, and corresponding molecular-dynamics simulations have been published [19, pp. 227–239]. We prefer a simpler method, devised by Berendsen et al. [26]. At each time step the instantaneous temperature is calculated and compared with T^0. All atomic velocities are then multiplied by

$$\left\{1 + \frac{\Delta t}{\tau_T}\left(\frac{T^0}{T} - 1\right)\right\}^{1/2} \tag{55}$$

where τ_T is an adjustable parameter, the relaxation time constant. If this parameter is infinitely large nothing is changed, whereas for $\tau_T = \Delta t$ the desired temperature would be reached instantaneously. Of course, this would catastrophically disrupt the system, but for a large value of τ_T the system can be considered as weakly coupled to a temperature bath.

Pressure can be regulated in the same way. After each time step in the simulation all coordinates are rescaled by the transformation $\mathbf{r}'_k = \mathbf{r}_k + \mathbf{D}\mathbf{r}_k$ with

$$\mathbf{D} = \frac{\beta \Delta t}{3\tau_P}(\mathbf{P} - P^0\mathbf{I}) \tag{56}$$

where β is the isothermal compressibility and τ_P is a pressure relaxation time. It can be shown that in the end the pressure components will converge to $P_{\alpha\beta} = P^0\delta_{\alpha\beta}$; in the isotropic case this would occur instantaneously for $\tau_P = \Delta t$. However, to avoid instabilities in the simulation it is important that this time constant be chosen quite large, at least 1 ps.

Of course, the coordinate scaling also applies to the axes of the simulation box (normally small multiples of the crystallographic cell constants **a**, **b**, **c**). Application of Eq. 56 may result in a rotation of the whole system which will destroy the translational symmetry assumed for the periodic boundary conditions. So a compensating rotation **R** should be performed after the scaling:

$$\mathbf{r}''_k = \mathbf{R}\mathbf{r}'_k = \mathbf{R}(\mathbf{r}_k + \mathbf{D}\mathbf{r}_k) = (\mathbf{R} + \mathbf{D})\mathbf{r}_k \tag{57}$$

where we have used the fact that the elements of **D** are much smaller than unity and, correspondingly, **R** can be expressed as the unit matrix plus a small antisymmetric matrix. To preserve the desired conditions $a_y = a_z = b_z = 0$ the matrix $\mathbf{R} + \mathbf{D}$ should be upper triangular, which forces the choice of **R** and leads to

$$\mathbf{R} + \mathbf{D} = \mathbf{I} + \begin{pmatrix} D_{xx} & D_{xy} + D_{yx} & D_{xz} + D_{zx} \\ 0 & D_{yy} & D_{yz} + D_{zy} \\ 0 & 0 & D_{zz} \end{pmatrix} \tag{58}$$

The physical justification of the scaling procedure is disputable. But this is not really relevant: after an equilibration period (a small multiple of the relaxation time) stable values of T^0 and P^0 should have been attained, so that the difference from a system at constant energy and volume will become negligible.

2.3.2 *Crystal Simulations*

Some remarks pertaining especially to crystal simulations are in order. Translational, and possibly other, symmetry elements should not be used as constraints for the trajectories. In this way unjustified limitations on atomic movements are avoided; moreover, at the end of the simulation it is essential to assess whether or not the required symmetry has been maintained. If not, the force field and/or the algorithm are inadequate, or a real possible disorder in the crystal must be envisaged. The calculation proceeds as follows: the vibrational amplitudes

$$\langle \Delta r_{k\alpha} \Delta r_{k\beta} \rangle = \langle (r_{k\alpha} - \langle r_{k\alpha} \rangle)(r_{k\beta} - \langle r_{k\beta} \rangle) \rangle \tag{59}$$

are calculated by averaging over the time steps of the simulation. Then the crystal symmetry operations are applied to all atomic coordinates to reduce them to one selected asymmetric unit, and the process is repeated. If the vibrational amplitudes are about the same as before, the system models a well-behaved crystal and the results of Eq. 59 can be compared with X-ray results. Here the vibrational amplitudes are expressed in a form that does not depend on the Cartesian axes system, usually either $\boldsymbol{\beta} = 2\pi^2 \mathbf{C}^* \langle \Delta \mathbf{r} \Delta \mathbf{r}^T \rangle (\mathbf{C}^*)^T$ [5] or $\mathbf{U}$ which is defined by relations like $U_{ab} = \beta_{ab}/(2\pi^2 a^* b^*)$ [1]. (In Ref. [2, p. 186] the symbols $\boldsymbol{\beta}$ and $\mathbf{U}$ are used differently, and the dimensions of various quantities entering in the discussion are not always clear.) Often only isotropic temperature factors are published; these are defined by

$$U_k^{\text{iso}} = \tfrac{1}{3} \sum_{\alpha} \langle \Delta r_{k\alpha}^2 \rangle \tag{60}$$

and may be also reported as B_k which is equivalent to $8\pi^2 U_k^{\text{iso}}$.

Vibrational amplitudes can also be compared with the results of a lattice-dynamical calculation, Eq. 42. Then one realizes that the computational box used in the molecular dynamics calculation is relatively small. More specifically, N_1, N_2, N_3 in Eq. 36 will be small integers (seldom exceeding, say, 4), so the calculation of vibrational amplitudes is equivalent to a lattice-dynamical calculation with a very limited sampling of $\mathbf{q}$ vectors. To the knowledge of the authors this theoretical sensitivity of the results to the size of the computational box has not yet been thoroughly investigated.

In principle it should be easy to find expansion coefficients by repeating the simulation at different temperatures. Elastic constants could be obtained in the same way by varying the external pressure anisotropically. In both cases high accuracy would be required since small differences in $\mathbf{C}$ are needed, and such calculations are still rare. Alternatively, one could accumulate the averages of the corresponding theoretical expressions [6], at the loss of the appealing simplicity of the simulational approach. Derivatives with respect to the temperature, as occurring in thermal-expansion coefficients and specific heats, are found from Eq. 18:

$$\frac{\partial\langle X\rangle}{\partial T} = \frac{1}{kT^2}(\langle XU\rangle - \langle X\rangle\langle U\rangle) \tag{61}$$

a fluctuation property which can hardly be expected to be more accurate than the direct approach [19]. The elastic constants are essentially derivatives of $P_{\alpha\beta}$ with respect to $C_{\gamma g}$; a very short introduction to the theory of elasticity is given in the Appendix. So one needs derivatives like

$$\frac{\partial\langle X\rangle}{\partial C} = \left\langle\frac{\partial X}{\partial C}\right\rangle + \frac{1}{kT}\left(\langle X\rangle\left\langle\frac{\partial U}{\partial C}\right\rangle - \left\langle X\frac{\partial U}{\partial C}\right\rangle\right) \tag{62}$$

Again, few such calculations have been reported.

It would be very desirable if *free energies* could be obtained from a molecular-dynamics calculation. One route is via Eq. 19 which can be developed into an interesting relation between the determinant of the $3N \times 3N$ matrix $\langle\Delta r_{k\alpha}\Delta r_{k'\beta}\rangle$ and the entropy S [27]:

$$S = \tfrac{1}{2}k\left(6N + \ln|\det\langle\Delta r_{k\alpha}\Delta r_{k'\beta}\rangle| + 3\sum_{k=1}^{N}\ln\frac{4\pi^2 m_k kT}{h^2}\right) \tag{63}$$

Harmonicity is assumed in this equation; a more general treatment, including the use of constraints, is given in Ref. [28].

Another procedure is to use thermodynamic perturbation or integration methods to transform the system to a reference system with known free energy. Briefly, the free-energy difference between two systems can be found by performing a set of simulations for hypothetical intermediate states. For the general theory the first two chapters of Ref. [29] are recommended (but note that the 'slow growth' method mentioned there has later turned out to be unreliable [30]). An obvious choice for the reference system is an Einstein crystal defined by:

$$U = \sum_{k=1}^{N} 2\pi^2 m_k \nu_k^2(\mathbf{r}_k - \mathbf{r}_k^{\mathrm{e}})^2 \tag{64}$$

where the ν_k can be arbitrarily chosen and inserted into Eq. 40 to give the free energy. This model is quite unrealistic (it does not depend on interatomic distances, so there is no dependence of the frequencies on the wave vector $\mathbf{q}$) but that does not matter since any reference system will do as long as classical dynamics can be applied. But here is the catch: in practice that treatment is only valid for low frequencies. For one thing, from the classical free energy of the Einstein crystal,

$$A = 3kT \sum_{k=1}^{N} \ln(h\nu_k/kT) \tag{65}$$

unphysical negative entropies are found for wave numbers as low as 600 cm^{-1}. So preferably high frequencies should be eliminated, for instance by considering the molecules to be rigid [31].

For phase transitions within the solid state only free-energy *differences* are needed, and these can be obtained more reliably. Direct simulations of such phase transitions have also been reported, see e.g. Refs [32] and [33] for sulphur hexafluoride.

2.3.3 Velocity Autocorrelation

It is of interest to calculate the *autocorrelation function* of the velocity of atom k:

$$C_k^{\mathrm{v}}(t) \equiv \langle \mathbf{v}_k(t_0 + t) \cdot \mathbf{v}_k(t_0) \rangle \tag{66}$$

where the averaging is done over all possible times t_0. For harmonic vibrations it can be shown [11, 22] that the frequency distribution $g(\nu)$ can be found from the Fourier transform of the velocity autocorrelation function:

$$g(\nu) = \frac{4}{kT} \sum_{k=1}^{N} m_k \int_0^{\infty} C_k^{\mathrm{v}}(t)\, \mathrm{e}^{-2\pi \mathrm{i} \nu t}\, \mathrm{d}t \tag{67}$$

This equation provides a very direct way from the simulation results to the frequency distribution function. Of course, what has been said above on the limited number of $\mathbf{q}$ vectors remains valid here.

2.4 Monte Carlo Methods

In the Monte Carlo method subsequent configurations of the system are generated by a random procedure rather than by the deterministic Newton equations. The energy U is calculated as usual. Then an atom is picked at random and displaced in an arbitrary direction over a fraction R of a certain preset maximum displacement, where R is a randomly chosen value between 0 and 1. Let the energy for this trial configuration be U'. Then the random move is accepted if

$$\mathrm{e}^{-(U'-U)/kT} > R' \tag{68}$$

where R' is another random number between 0 and 1. If U' is much larger than U it is very unlikely that the move will be accepted; for $U' < U$ acceptance is a certainty. As only an energy *difference* is needed the computational effort can be quite limited.

After each move a new configuration is recorded: if the move was accepted this must be the trial configuration, otherwise the unchanged starting configuration is copied. This procedure is repeated a great number of times. It can be shown that the set of configurations obtained in this way is representative of the actual system, so expectation values for observable properties can be obtained by simple averaging. It must be noted that the random moves have no connection with trajectories in time, so no time-dependent properties can be obtained.

Only the most simple form of the Monte Carlo method, suitable for monoatomic substances, has been described here. A trial movement of one atom in a molecule is almost certain to increase the energy considerably and has, therefore, a very limited probability of being accepted. Random translations and rotations of an entire molecule have a better chance. For nonrigid molecules sampling can best be done by random changes in selected internal coordinates, which entails some theoretical complications as indicated in the discussion on constraints in Section 2.3. Therefore the molecular dynamics method is generally easier to apply for complex molecules.

3. FORCE FIELDS

3.1 Parametrization of Force Fields

The construction of a force field has two aspects: the first is the choice of the functional form into which Eq. 1 shall be cast, and the second is the choice of the parameters that inevitably occur in it. In principle both should be done for every molecular system, but this is clearly infeasible. Fortunately, it turns out that quite acceptable results can be obtained if not only the functional form but even many of the parameters are taken to be transferable between molecules within a certain chemical class. This is in line with the common chemical knowledge that standard bond lengths, bond energies and so on *are* useful concepts. Without such transferability the benefit of molecular modelling would be very limited indeed.

Early force fields (of the 1960s and 1970s) were parametrized against experimental gas-phase data, including vibrational frequencies, geometries and rotational barriers from electron diffraction and microwave spectra, heats of formations, excess enthalpies, etc. The MM2/MM3 force fields [34, 35], the Ermer force field [36] and CFF [37] find their roots in these developments. Initially the simultaneous requirement of reproducing structural and spectroscopic data could not be achieved. Therefore separate force fields were designed to focus on each of these aspects. Only *consistent force fields* [37, 38] aimed at doing both. Parallel to these developments, which were mainly focussed on small organic molecules, several simple potential-energy functions were being used for peptides [39, 40]. In the 1980s three major protein/nucleic acid force fields emerged and became widely used: CHARMM [41], AMBER [42] and GROMOS [43]. These force fields are mainly parametrized against crystal structure data. In addition the ECEPP/UNICEPP force fields [44, 45], which consider only rotational degrees of freedom, are worth mentioning. Currently, several commercial packages contain their own force fields like CVFF (Discover by Biosym), or SYBYL (Tripos) or some version of AMBER (Discover by Biosym) or CHARMM (CHARMm by Molecular Simulations).

For a detailed description of organic molecules the MM3 force field is now the state of the art; it is also used for molecular crystals [46]. MM2 was developed on gas-phase data for molecules with few polar groups, and the electrostatics and hydrogen-bonding interactions were not always properly described. This has been largely corrected in MM3. It has a rather complex functional form with many coupling terms (like stretch–bend, torsion–stretch and torsion–bend terms), anharmonicity terms and many different atom types, and incorporates subtle corrections such as that for the anomeric effect in carbohydrates. It has been refined to such a level that it now also reproduces vibrational frequencies quite well.

In line with early work [47] the Hagler group systematically searches for the proper functional forms of the potentials based on *ab initio* calculations. They arrived at a functional form in their CFF91 force field whose complexity is comparable to that of MM3 [48].

For proteins it has been common practice to use simple force fields, because the size of the molecular system would otherwise prohibit extensive MD simulations. CHARMM, AMBER and GROMOS all used to be united atom force fields (see below) and to have a simple functional form (diagonal, i.e. no cross terms in the potentials).

We will now describe some aspects of parametrization and some differences between the various force fields to get a feeling for the significance of the particular descriptions of the individual terms in the interactions.

3.1.1 General Form

Most force fields contain harmonic terms for bond stretching and for valence angle bending, and Fourier terms for internal rotation. But an essential part of any force field has turned out to be that atom–atom potentials are also applied to intramolecular pairs of nonbonded atoms. So a very basic force field would look like this:

$$U = \sum_{\text{bonds}} \tfrac{1}{2}K_b(b-b_0)^2 + \sum_{\text{angles}} \tfrac{1}{2}K_0(\theta-\theta_0)^2 + \sum_{\text{dihedrals}} \sum_{n} K_{n\phi}\cos(n\phi)$$
$$+ \sum_{k} \sum_{k'>k} {}' U^{\text{vdw}}_{kk'}(l_{kk'}) + \sum_{k} \sum_{k'>k} {}' U^{\text{el}}_{kk'}(l_{kk'}) \qquad (69)$$

The primes in the last summations denote that it is customary to omit nonbonded interactions between atoms connected via one or two bonds, as these interactions are already modelled by the harmonic potentials. The $U^{\text{vdw}}_{kk'}$ and $U^{\text{el}}_{kk'}$ denote the *van der Waals* and electrostatic pair potentials (see below). Sometimes a phase angle is added to the cosine argument $n\phi$ in the dihedral contribution; only 0 and π are permitted by symmetry, and it is better to omit the phase angle and to allow negative values of $K_{n\phi}$.

It is clear that Eq. 69 contains more energy terms than there are degrees of freedom in the system being studied. So one must not make the mistake of considering, for instance, b_0 as an equilibrium bond length and K_b as its force constant. Equilibrium bond lengths must be found by finding the energy minimum of the entire force field, and the force constants by taking the second derivatives of U at that minimum. Likewise, the torsional parameters $K_{n\phi}$ cannot be compared with the Fourier components of internal rotation barriers determined by microwave spectroscopy on free molecules: the latter refer to the total energy where nonbonded interactions play an essential role.

As is evident from Eq. 69, the energy is not on an absolute scale: differences between the results for different molecules need not give reaction energies. The term *steric energy* is often used for that reason. It should be useful to calculate properties for each separate substance. Even then, depending on the way the force field was obtained, some properties are often better reproduced than others. For instance, it is clear that a force field optimized only on geometrical data would contain an arbitrary scale factor for the energy and could never produce thermodynamical data. More generally, without using data on torsional barriers in free molecules only the low-energy regions of configurational space are available for the parametrization since the high energy regions are hardly visited. Obviously, calorimetric data are sensitive to the potential energy function itself, geometrical data to its first derivatives and spectroscopic data to its second derivatives. As mentioned before, ideally a consistent force field should reproduce all such observations, but few force fields are up to that standard.

3.1.2 Van der Waals Nonbonded Interactions

Nonbonded interactions in or between molecules are broken down into atom–atom potentials. Assuming that atoms have a more or less spherical distribution of electrons around them, a simple interaction, reminiscent of that of noble gases, is postulated. Two basic functional forms are in use. The *Lennard-Jones* potential has both the repulsive and the attractive term as powers of the inverse distance between atoms:

$$U^{\text{vdw}}_{kk'}(l_{kk'}) = \frac{n\varepsilon^*}{n-m}\left[\frac{m}{n}\left(\frac{r^*}{l_{kk'}}\right)^n - \left(\frac{r^*}{l_{kk'}}\right)^m\right] \tag{70}$$

The most frequently used exponents are $n = 12$ and $m = 6$, in which case the van der Waals interaction is called a 12-6 potential. This potential is used in two different versions, the first one being:

$$U_{kk'}^{\text{vdw}}(l_{kk'}) = \frac{\varepsilon^* r^{*12}}{l_{kk'}^{12}} - \frac{2\varepsilon^* r^{*6}}{l_{kk'}^{6}} = \frac{A_{kk'}}{l_{kk'}^{12}} - \frac{B_{kk'}}{l_{kk'}^{6}} \tag{71}$$

The expressions are equivalent but the combination rules are different. The parameters ε^* and r^* are dependent on the type of atoms involved and are generally found from atomic properties ε_k and r_k by $\varepsilon^* = \sqrt{\varepsilon_k \varepsilon_{k'}}$ and $r^* = r_k + r_{k'}$, whereas for the right-hand expression $A_{kk'} = \sqrt{A_k A_{k'}}$ and similarly for B. For a detailed discussion of combination rules see [49]. The minimum in the potential occurs at $l = r^*$ with a depth of ε^*. The second version is:

$$U_{kk'}^{\text{vdw}}(l_{kk'}) = \frac{4\varepsilon^* \sigma^{*12}}{l_{kk'}^{12}} - \frac{4\varepsilon^* \sigma^{*6}}{l_{kk'}^{6}} \tag{72}$$

The minimum occurs at $l = \sqrt[6]{2}\sigma^*$ with a well depth of ε^*.

The second functional form is the *Buckingham* potential which has three adjustable parameters instead of two:

$$U_{kk'}^{\text{vdw}}(l_{kk'}) = \frac{\varepsilon^*}{1 - (6/\alpha)} \left[\frac{6}{\alpha}\, \mathrm{e}^{\alpha(1 - l_{kk'}/r^*)} - \left(\frac{r^*}{l_{kk'}} \right)^6 \right] \tag{73}$$

This function has been modified to a two-parameter form by Hill:

$$U_{kk'}^{\text{vdw}}(l_{kk'}) = c_1 \varepsilon^*\, \mathrm{e}^{-c_2 l_{kk'}/r^*} - c_3 \varepsilon^* \left(\frac{r^*}{l_{kk'}} \right)^6 \tag{74}$$

where the c's are predetermined constants, chosen in such a way that the minimum occurs at $l = r^*$ and with a depth of $1.17\varepsilon^*$. This form is used by MM2/MM3. The Buckingham potential has the drawback of going to $-\infty$ for $l \to 0$, so one must take precautions at very short distances.

So, although the parameters have the same symbols, they have a slightly different meaning. This makes them less transferable between the various force fields.

Lifson and Warshel [38] set out to determine which of the functional forms works best in reproducing data on organic and peptide molecules. They found no significant difference between Buckingham exp-6 and Lennard-Jones 12-6. Nevertheless it is felt that the exponential part gives a more realistic description of the exchange repulsion, as it goes up less steeply with decreasing distances below the van der Waals radius. This was also the reason for investigating the 9-6 Lennard-Jones potential, but no clear improvement was found [47], probably because most distances in the structural data are near the van der Waals radius. There is not much discussion on the attractive l^{-6} part of the potential which represents the induction and dispersion interactions. An additional computational advantage of the 12-6 potential in simulations of large molecular systems is that once l^{-6} is calculated, l^{-12} is just its square.

To reduce the computational demands it has been common practice to reduce the number of atoms by using united (extended) atoms, where the aliphatic hydrogens have become an integral part of their parent carbon atoms. The van der Waals radius is then significantly extended. The recently developed CHEAT [50], an adaptation of the CHARMM force field, uses united polar groups.

Related to this functional form is the use of special hydrogen-bond potentials and scaling of the so-called 1,4-interactions.

3.1.3 1,4 Nonbonded Interactions

The energy dependence on the dihedrals is described by torsional potentials as far as it is not governed by the 1,4 nonbonded interactions. It turned out that the combined use of van der Waals parameters for normal as well as for 1,4 nonbonded interactions was rather problematic in the case of the Lennard-Jones potential. This is because the l^{-12} repulsion curve is too steep, as becomes especially noticeable for close 1,4-interactions, and because charge redistribution in such cases has the effect of reducing the repulsion. This effect becomes even more problematic when united atom parameters are used. Therefore either separate 1,4 van der Waals parameters (CHARMM, GROMOS) or a scaling factor, usually of 0.5 (AMBER), should be applied. Sometimes also the electrostatic interactions are scaled down. Due to the fact that MM2/MM3 uses the $\exp(-l)$ potential, no such special treatment is necessary there.

3.1.4 Electrostatics

Usually each atom is assigned a net partial charge q_k from which the Coulomb nonbonded interactions are calculated:

$$U^{\mathrm{el}}_{kk'}(l_{kk'}) = \frac{q_k q_{k'}}{4\pi\epsilon_0 \epsilon l_{kk'}} \tag{75}$$

where ϵ is the relative dielectric constant. The charges can be obtained in various ways. In GROMOS, CHARMM and CVFF these are integral parts of the force field, and are fitted to experimental structural data. The Kollman group [51] has the philosophy that charges represent the electrostatic behaviour realistically if they are fitted to the *ab initio* molecular potential for each molecule (or part of it) to be simulated. They prescribe the level at which this *ab initio* calculation has to be carried out to be compatible with the rest of the force field. In MM2/MM3 the electrostatics is based on bond dipoles that are chosen so as to reproduce experimental dipole moments of small molecules in the gas phase. The electrostatic energy is expressed as:

$$U_{\mathrm{AB}}^{\mathrm{el}} = \frac{1}{4\pi\epsilon_0\epsilon}\left[\frac{\boldsymbol{\mu}_{\mathrm{A}}\cdot\boldsymbol{\mu}_{\mathrm{B}}}{l_{\mathrm{AB}}^3} - 3\frac{(\boldsymbol{\mu}_{\mathrm{A}}\cdot\mathbf{r}_{\mathrm{AB}})(\boldsymbol{\mu}_{\mathrm{B}}\cdot\mathbf{r}_{\mathrm{AB}})}{l_{\mathrm{AB}}^5}\right] \tag{76}$$

where μ is the bond dipole, $\mathbf{r}_{\mathrm{AB}}$ is the vector between the midpoints of the bonds A and B, and l_{AB} is its modulus.

The relative dielectric constant ϵ that occurs in these formulae needs some comment. Strictly speaking for microscopic interactions its value should be 1. A different dielectric constant is often used for the following reasons: (i) dipoles within the molecule become polarized by their mutual presence, which is basically an attractive interaction, but distance-dependent. This argument has been used to choose $\epsilon = l_{kk'}$ (note the tacit assumption of Angstrom units!); (ii) to compensate for the lack of solvent in certain simulations. Dielectric-solvent molecules reduce the electric field of one dipole at the position of another dipole within the molecule, by their reorientation. This effect is also distance-dependent [52]. Therefore $\epsilon = cl_{kk'}$ with $c > 1$ Å^{-1} is chosen.

If all (solvent) molecules are included in the calculation, like in crystal-structure simulations, then $\epsilon = 1$ is the proper choice. It can be argued that polarization effects (i) are not accounted for but it has to be realized that often polar groups are parametrized as if in a polarized state. For instance, in water the atomic charges are chosen to reproduce the dipole moment in liquid water (about 2.4 D) rather than in the gaseous state (1.85 D). Clearly an approach where polarization is included in the calculations explicitly is to be preferred (see Section 3.3.2).

3.1.5 Hydrogen-Bond Potential

Hydrogen bonds would be expected to be treated adequately by the use of an exp-6 or 12-6 potential plus electrostatics because the main interactions, Coulomb, exchange repulsion, induction and dispersion attraction, are accounted for. Nevertheless this turned out to be a major problem, because the hydrogen in an A—H...B hydrogen bond finds itself well within the van der Waals radius of B and therefore outside the range fitted in the van der Waals potentials. To obtain accurate hydrogen-bond distances and geometries two approaches are followed. (i) A separate hydrogen-bond potential for the H...B or A...B interaction is used. This could be either a Morse [47, 53] or a 12-10 Lennard-Jones potential (first versions of AMBER), or a Buckingham potential with separate parameters (MM3), sometimes attenuated by an angle-dependent term [54]. (ii) The van der Waals interaction of polar H's is set to zero (GROMOS). The latter approach works, but the polar A...B Lennard-Jones repulsions have to be increased.

In our view no special hydrogen-bond potential would be needed if the van der Waals potential of especially polar hydrogens is sufficiently soft (see also the discussion in Chapter 3, Section 1.2.1).

3.1.6 Final Remarks

A review of recent literature is beyond the scope of this chapter. An excellent review is given by Halgren [55]. It will suffice here to mention some newly developed force fields. The CFF93 force field of the Hagler group [56, 57] is a so-called QMFF (quantum-mechanical force field) or class II force field. The functional form and the initial parameters are derived from *ab initio* potential-energy surfaces, at the HF 6-31G* level. The parameters are finally fitted in an automated procedure to a large collection of experimental data. A similar approach is used by Halgren [58] in the development of the Merck MMFF94 force field, but this author used higher-level quantum chemistry (TZP basis set and MP4 perturbation theory).

Newly parametrized AMBER [59] and CHARMM [60] force fields are emerging: the so-called second-generation force fields. AMBER is now based on RESP (restrained electrostatic potential) atomic partial charges. These are obtained by a restricted fit to *ab initio* electrostatic potentials at the HF 6-31G* level, and the other parameters are mainly fitted to solution structural data. The new CHARMM force field employs fits to quantum-mechanical intermolecular interaction energies.

Throughout, specializations of the well-known force fields for specific classes of molecules (e.g. carbohydrates) have appeared. Hopefully, with the improvement of functional forms and inclusion of polarization effects this will no longer be necessary in future. Models that include polarizability are beginning to appear, but till now mainly for simple liquids.

The various differences in functional forms suggest that there are several ways to handle the same type of interactions. A sophisticated force field may contain so many parameters that the results are hardly changed when certain combinations of parameters, e.g. torsional potentials and nonbonded interactions, are varied in the right way. It follows that there are no unique or physically meaningful values for the parameters, and that it is not possible to transfer seemingly comparable parameters from one force field to the other [61].

When evaluating results from force field calculations there is always the danger of being impressed by their complexity and, in modern program packages, by their beautiful presentation. It is important to keep in mind all the assumptions that have been made in developing the underlying force field.

3.2 Crystals

A natural description of molecular crystals considers the molecules as rigid units, so only intermolecular interactions have to be taken into account. For all but the smallest molecules this is usually done by assuming that these interactions are the sum of atom–atom or van der Waals potentials. In the pioneering work of Kitaigorodsky [6] a geometric model was used: hard spheres, corresponding to $U_{kk'}^{\mathrm{vdw}}$ either zero or infinity. Even this extremely simple model proved to be valuable, e.g. in understanding the preponderance of a few space groups over most others. After the introduction of Lennard-Jones or Buckingham potentials, the atom–atom approach can even give a quantitative description of many physical properties, as shown in the early work on hydrocarbons by Kitaigorodsky [6] and Williams [62].

A noteworthy observation is made when molecules are neighbours in a crystal. Most atoms will be separated so far that their nonbonded interaction is attractive. So the distance between two atoms where the molecules actually touch (called by crystallographers the *van der Waals distance*) is considerably shorter than the minimum in their pair-interaction curve, to reach an equilibrium between attractive and repulsive interactions. For instance, the H–H van der Waals distance in crystals is about 2.4 Å whereas the pair-interaction curve has a minimum at almost 3.0 Å. One is led to suspect that, counterintuitively, very short intermolecular contacts may be repulsive rather than attractive.

A better description of a real crystal can be obtained if intramolecular degrees of freedom are taken into account. Bond lengths will be quite insensitive to the crystal environment, but bond angles are already somewhat more variable and dihedral angles even more so. For instance, hydrogen bonds between carbohydrate molecules are possible only if the hydroxyl groups are favourably oriented after internal rotation about the C–O bonds.

Pertsin and Kitaigorodsky [4, p. 79] have given an extensive review of procedures to derive force-field parameters for crystals. Usually crystal data are used to derive parameters for intermolecular interactions and data on free molecules for intramolecular parameters.

In Section 4.3 a very demanding goal will be set: not only should a force field reproduce the observed crystal structure well, but also it should be shown that this structure is the global rather than just a local minimum. In other words, the force field should be able to predict the observed crystal structure(s) out of the many conceivable ones.

3.3 Calculation of Energy, Forces and Pressure

The energy is always calculated for N independent atoms. In the case of molecular mechanics all symmetry-related atoms are equivalent, and N is the number of atoms in the asymmetric unit. In molecular dynamics N is equal to the number of atoms in the computational box since no symmetry is imposed within that box. In both cases the number of intramolecular energy terms is limited, and almost all effort is spent on the calculation of the nonbonded interactions whose number is, in principle, infinite:

$$U^{\text{nonb}} = \tfrac{1}{2} \sum_{k=1}^{N} \sum_{i=1}^{\infty} \sum_{k'=1}^{N} u_{k,ik'}(l_{k,ik'}) \tag{77}$$

where i numbers the replicas of the independent atoms that are created by symmetry operations (Eq. 11), including the lattice translations; $l_{k,ik'}$ denotes the distance between an atom k and an atom k' which may be in the same asymmetric unit ($i = 1$) or in a symmetry-related set ($i > 1$). Obviously, $u_{k,1k} = 0$.

In practice, of course, only those terms that contribute significantly are retained. Usually this is done by introducing a maximum interatomic distance, the *cutoff* radius, of the order of 10 Å. For the short-ranged repulsive interactions this introduces a negligible error, but for the l^{-6} dispersion terms the cutoff error may already become important. For Coulomb interactions, with their slow decline proportional to l^{-1}, the errors can become intolerably large. To alleviate this problem the molecule may be subdivided into *charge groups* with zero or small total charge, and the cutoff criterion is then based on the centres of these charge groups. In this way the neglected interactions are of dipole–dipole type, and their decay is proportional to l^{-3}.

More sophisticated methods to treat the long-range Coulomb and dispersion terms are discussed in Section 3.3.1. The aim is to eliminate cutoff effects. However, it is important to realize that this does not necessarily improve the performance of a force field that was developed with a certain cutoff radius; one might argue that the choice of a cutoff radius is an essential component of a force field like any other parameter.

The forces on the atoms are calculated from:

$$\mathbf{F}_k^{\text{nonb}} = -\sum_{i=1}^{\infty} \sum_{k'=1}^{N} \frac{\partial u_{k,ik'}}{\partial l_{k,ik'}} \frac{\mathbf{r}_k - \mathbf{r}_{ik'}}{l_{k,ik'}} \tag{78}$$

In molecular dynamics the pressure can be calculated from Eq. 28. Here special care has to be taken when bond lengths are kept fixed in the simulation because then constraint forces contribute to the virial [23] in Eq. 31. It can be shown that the contribution of bond angles and torsional angles is always zero for isotropic systems [63, 64], but for crystals this is not the case. For small molecules the formulation of Eq. 32 is preferable since it contains no intramolecular contributions, but for a polymeric structure Eq. 28 must be used.

Correspondingly, in molecular mechanics the derivative of the energy to some parameter (p) may be desired. This can be obtained by straightforward differentiation of Eq. 77:

$$\frac{\partial U^{\text{nonb}}}{\partial p} = \tfrac{1}{2} \sum_{k=1}^{N} \sum_{i=1}^{\infty} \sum_{k'=1}^{N} \frac{\partial u_{k,ik'}}{\partial l_{k,ik'}} \frac{\mathbf{r}_k - \mathbf{r}_{ik'}}{l_{k,ik'}} \cdot \frac{\partial(\mathbf{r}_k - \mathbf{r}_{ik'})}{\partial p} \tag{79}$$

For instance, if p is one of the cell parameters, the derivative of the coordinate

$r_{ik'\beta}$ with respect to $C_{\alpha g}$ is $\delta_{\alpha\beta}t_{ig}$, as follows from Eq. 11. Alternatively, one can prefer to write the energy as a function of fractional coordinates and obtain a different expression. If the molecules in the crystal are considered as rigid units, a standard rigid geometry $\mathbf{r}_k^{\text{ref}}$ must be known. The positions in the asymmetric unit are then defined by a translational vector **T** and a rotational matrix **R**:

$$\mathbf{r}_k = \mathbf{T} + \mathbf{R}\mathbf{r}_k^{\text{ref}} \tag{80}$$

and Eq. 79 can be used where p now is one of these six parameters. For detailed equations see Williams [65] or Pertsin and Kitaigorodsky [4], where expressions for second derivatives can be found too. The latter authors also give a detailed account of the calculation of the elements of the dynamical matrix (Eq. 37) where the Cartesian coordinates are replaced by internal coordinates, and review some typical applications of lattice-dynamical calculations.

In the calculations it is possible to save some computer time by limiting the summations to $k' \geqslant k$, carefully observing the symmetry relations in the space group involved. The details are tedious and shall not be given here.

A part of the computational effort is used to determine whether or not a certain interatomic distance falls within the cutoff radius. For a crystal, where the deviations from the equilibrium position are limited, this has to be done only once. So it is a good idea to construct a *pair list* for use at following time steps in molecular dynamics or iterations in energy minimization. Nevertheless, important structural changes might occur and the pair list should be updated from time to time.

3.3.1 *Convergence Acceleration*

In the calculation of Coulomb and dispersion terms in the nonbonded energy, Eq. 77, the following summations occur:

$$E_n = \tfrac{1}{2}\sum_{k=1}^{N}\sum_{i=1}^{\infty}\sum_{k'=1}^{N}{}' B_{kk'} l_{k,ik'}^{-n} \tag{81}$$

where the B's are force-field parameters. In this context N will be the number of atoms in a unit cell with volume V. The symmetry operations indicated by i can now only be translations to an equivalent cell; as before, $i = 1$ means no operation. The prime signifies that the terms $i = 1$, $k = k'$ must be excluded from the summations.

For $n = 6$ we have the dispersion energy E_6 which converges reasonably fast, yet a cutoff distance of 10 Å may be insufficient for accurate work. Although each individual term beyond that distance is small, there are many contributions and they are all attractive. A very simple correction is to assume a homogeneous distribution of atoms outside the cutoff radius R_c, so there is one interaction of the type $k - k'$ in a volume V and in Eq. 81 the neglected contribution is approximately:

$$E_n^{\text{corr}} = \frac{1}{2}\sum_{k=1}^{N}\sum_{k'=1}^{N}\int_{l=R_c}^{\infty}\frac{B_{kk'}}{l^n}\frac{1}{V}4\pi l^2\,\mathrm{d}l = \sum_{k=1}^{N}\sum_{k'=1}^{N}\frac{2\pi B_{kk'}}{(n-3)VR_c^{n-3}} \tag{82}$$

For $n = 1$ we have the Coulomb energy which converges only conditionally. There are positive as well as negative contributions and E_1 fluctuates widely as a function of the cutoff distance. Especially in polar space groups the convergence of the series is bad [66]. Around 1920 Madelung and Ewald devised efficient methods to calculate energies of ionic lattices. Their ideas were generalized to any n by Nijboer and de Wette [67] and used by Williams [68] to calculate the lattice energy of benzene. The result can be expressed as:

$$E_n = \frac{1}{2\Gamma(n/2,0)}\left\{\sum_{k=1}^{N}\sum_{i=1}^{\infty}\sum_{k'=1}^{N}{}' B_{kk'}l_{k,ik'}^{-n}\Gamma(n/2, l_{k,ik'}^2/\eta^2) - \frac{2}{n\eta^n}\sum_{k=1}^{N}B_{kk}\right.$$
$$\left. + \frac{\pi^{n-3/2}}{V}\sum_{\mathbf{H}\neq 0}A_{\mathbf{H}}H^{n-3}\Gamma\left(\frac{3-n}{2}, \pi^2H^2\eta^2\right) + \frac{2A_0\pi^{3/2}}{(n-3)V\eta^{n-3}}\right\} \tag{83}$$

where $\Gamma(a, b)$ is the incomplete gamma function [15]:

$$\Gamma(a, b) \equiv \int_{x=b}^{\infty} \mathrm{e}^{-x}x^{a-1}\,\mathrm{d}x \tag{84}$$

The parameter η can be chosen at will, $\mathbf{H}$ is a vector in reciprocal space:

$$\mathbf{H} = h_1\mathbf{a}^* + h_2\mathbf{b}^* + h_3\mathbf{c}^* \tag{85}$$

with integer h_1, h_2, h_3, and $A_{\mathbf{H}}$ is defined as:

$$A_{\mathbf{H}} = \sum_{k=1}^{N}\sum_{k'=1}^{N}B_{kk'}\,\mathrm{e}^{2\pi\mathrm{i}\mathbf{H}\cdot(\mathbf{r}_k - \mathbf{r}_{k'})} \tag{86}$$

A very important simplification is possible if $B_{kk'}$ can be written as a product:

$$B_{kk'} = f_k f_{k'} \tag{87}$$

because then Eq. 86 reduces to

$$A_{\mathbf{H}} = \left|\sum_{k=1}^{N} f_k\,\mathrm{e}^{2\pi\mathrm{i}\mathbf{H}\cdot\mathbf{r}_k}\right|^2 \tag{88}$$

and the double summation over the atoms in the cell is reduced to a single one, comparable in form to the structure factor in diffraction theory.

Equation 83 is only valid for $n > 3$. Nevertheless, it is possible to find an expression for the Coulomb term E_1 since the unit cell has no net charge. It turns out that in this case Eq. 83 remains valid except for the last term which must be replaced by:

$$\frac{1}{6\epsilon_0 V}\left(\sum_{k=1}^{N} q_k \mathbf{r}_k\right)^2 \tag{89}$$

which is nonzero if the unit cell has a dipole moment. Then the crystal has a surface charge; if it is surrounded by a conducting medium this charge vanishes, and so does the term (89). For a detailed discussion of this subtle problem see Refs [19, 69]; the general practice is to omit the term (89).

The efficiency of the convergence acceleration can be optimized by a proper choice of η. The first term in Eq. 83 is a sum in real space, with the incomplete gamma-function as a damping factor. For small values of η the damping becomes large, and a small cutoff radius is possible. However, the third term in Eq. 83 is a summation in reciprocal space which converges quicker for large η. The optimum value of η must be found by trial and error; it is usually around 3 Å.

In principle this procedure should be applicable to the Coulomb interaction. However, in many force fields interactions between bonded atoms as well as 1–3 interactions are excluded, and the Coulomb energy cannot be cast in the form of Eq. 81. This complication can be easily handled [70] by considering the proper correction terms explicitly. The same situation arises in the dispersion energy, with the additional complication that not all force fields have parameters that follow the combination rule given by Eq. 87.

The forces are found by direct differentiation of E_n in Eq. 83. For the virial Eq. 31 cannot be used since the energy is no longer a function of the interatomic distances, and it is necessary to use Eq. 30. In that differentiation fractional coordinates must be considered as constant, and the same goes for $A_{\mathbf{H}}$. The equations tend to become complicated. Explicit expressions for energies, forces and virial components have been given by Karasawa and Goddard [71] for $n = 1$ and $n = 6$ (note that their $\mathbf{h}$ corresponds to our $2\pi\mathbf{H}$ and their $\mathbf{H}$ to our $\mathbf{C}$).

The 'Ewald summation' method summarized above is the most popular but by no means the only method to deal with Coulomb sums. We mention the related method devised by Bertaut (see Ref. [72] for references and further development). For a general review see Ref. [18] and, for recent developments, Ref. [73].

3.3.2 Polarization

A simple way to model polarization is to assume that a point dipole $\mathbf{p}_k$ can be induced at certain positions $\mathbf{r}_k$ that usually coincide with point atoms. The induced dipole is a function of the electric field strength, $\mathbf{p}_k = \mathbf{p}_k(\mathbf{E}_k)$. To create these dipoles a self-energy is needed [74]:

$$U^{\text{self}} = \sum_k \int_{\mathbf{p}'_k=0}^{\mathbf{p}_k} \mathbf{E}_k \cdot \mathrm{d}\mathbf{p}'_k = \sum_k \left(\mathbf{E}_k \cdot \mathbf{p}_k - \int_{\mathbf{E}'_k=0}^{\mathbf{E}_k} \mathbf{p}_k \cdot \mathrm{d}\mathbf{E}'_k \right) \tag{90}$$

The field strength at position k can be written as:

$$\mathbf{E}_k = \mathbf{E}_k^0 + \sum_{k'} \mathbf{T}_{kk'} \mathbf{p}_{k'} \tag{91}$$

where the first term gives the field from the permanent charges and the second term represents the contribution from the induced dipoles via the dipole field tensor **T**. The total polarization energy is then [75]:

$$U^{\text{pol}} = -\tfrac{1}{2} \sum_k \mathbf{p}_k \cdot (\mathbf{E}_k + \mathbf{E}_k^0) + U^{\text{self}} \tag{92}$$

with corresponding forces:

$$F_{k\alpha}^{\text{pol}} = \sum_{k'} \sum_{\beta} \frac{\partial E_{k'\beta}^0}{\partial r_{k\alpha}} p_{k'\beta} + \tfrac{1}{2} \sum_{k'} \sum_{k''} \sum_{\beta} \sum_{\gamma} p_{k'\beta} \frac{\partial T_{k'k''}^{\beta\gamma}}{\partial r_{k\alpha}} p_{k''\gamma} \tag{93}$$

As the induced dipoles depend on the atomic coordinates, the derivation of this equation is not straightforward.

In an ideal model of point charges one gets:

$$\mathbf{E}_k^0 = \frac{1}{4\pi\epsilon_0} \sum_{k' \neq k} \frac{q_{k'}(\mathbf{r}_k - \mathbf{r}_{k'})}{l_{k,k'}^3} \tag{94}$$

and

$$T_{kk'}^{\alpha\beta} = \frac{1}{4\pi\epsilon_0 l_{k,k'}^3} \left[\frac{3(r_{k\alpha} - r_{k'\alpha})(r_{k\beta} - r_{k'\beta})}{l_{k,k'}^2} - \delta_{\alpha\beta} \right] \tag{95}$$

and, after a tedious calculation [76], one can derive the expression for the forces:

$$\begin{aligned} \mathbf{F}_k^{\text{pol}} = \frac{1}{4\pi\epsilon_0} \sum_{k' \neq k} \Bigg[& \frac{q_{k'}\mathbf{p}_k - q_k \mathbf{p}_{k'}}{l_{k,k'}^3} + \frac{3}{l_{k,k'}^5} (q_k \mathbf{p}_{k'} \cdot \mathbf{r}_{k,k'} - q_{k'} \mathbf{p}_k \cdot \mathbf{r}_{k,k'}) \mathbf{r}_{k,k'} \\ & + \frac{3}{l_{k,k'}^5} \{ (\mathbf{p}_{k'} \cdot \mathbf{p}_k) \mathbf{r}_{k,k'} + (\mathbf{p}_{k'} \cdot \mathbf{r}_{k,k'}) \mathbf{p}_k + (\mathbf{p}_k \cdot \mathbf{r}_{k,k'}) \mathbf{p}_{k'} \} \\ & - \frac{15}{l_{k,k'}^7} (\mathbf{p}_{k'} \cdot \mathbf{r}_{k,k'})(\mathbf{p}_k \cdot \mathbf{r}_{k,k'}) \mathbf{r}_{k,k'} \Bigg] \end{aligned} \tag{96}$$

where $\mathbf{r}_{k,k'} \equiv \mathbf{r}_k - \mathbf{r}_{k'}$.

Usually the induced dipole is assumed to be a linear function of the field strength:

$$\mathbf{p}_k = \alpha_k \mathbf{E}_k \tag{97}$$

where α_k is the *polarizability* of atom k. Then the self-energy is easily calculated, and a very simple equation for the polarization energy is found:

$$U^{\text{pol}} = -\tfrac{1}{2}\sum_k \mathbf{p}_k \cdot \mathbf{E}_k^0 \qquad (98)$$

It is seen that the dipoles depend on field strength via Eq. 97 and vice versa via Eq. 91. This system of equations can be cast in linear form and solved for $\mathbf{E}_k$ and $\mathbf{p}_k$ by standard matrix methods [77, 78]. But this is quite laborious, and an iterative solution may be more practical.

An external field can be added to Eq. 94. For a free molecule it is then possible to calculate the total molecular polarizability tensor, and comparison with experimental results can lead to adjustment of the atomic polarizabilities α_k [77].

Unfortunately, sometimes the system is unstable and induced dipole moments tend to infinity. To avoid this catastrophe more sophisticated models have been introduced, where the form of $\mathbf{T}$ and/or $\mathbf{E}^0$ is modified by damping functions which become important at small interatomic distances [78, 79]. Such models have no very convincing background. One might perhaps just as well replace Eq. 97 by a nonlinear relation where the induced dipole moment cannot exceed a certain limiting value. It is also unclear how bonded atoms and 1–3 interactions should be treated. Although the number of studies using polarizable models (mostly on aqueous systems) is increasing rapidly, there is no general consensus on the best approach.

4. APPLICATIONS

In this section we shall mention some applications of molecular mechanics and molecular dynamics calculations, based mainly on our own work.

4.1 Validation of Crystal Structures

Experimental crystal structures are used to verify and improve force fields. The unit cell from an energy minimization should correspond to the experimental data within reasonable limits, say, a few percentage points in cell axes and a few degrees in cell angles. Moreover, the atoms within the cell should also be placed correctly. The comparison with experimental atomic positions can best be based on fractional coordinates, to eliminate the influence of errors in cell dimensions. Even then, this straightforward comparison will obscure many interesting details of the molecular structure. So it is preferable to distinguish between the position and orientation of the molecules in the cell on the one hand, and their internal structure on the other. Here one should consider bond lengths, bond angles, and especially dihedral angles.

In a molecular-dynamics simulation the effect of vibrations is included, so the results should depend on the temperature and, in principle, cannot be compared with those of an energy minimization. But the differences are usually smaller than the errors due to the force field. After the simulation the same comparisons with the experimental results should be made. Now also temperature factors can be compared, but here the experimental data may be less reliable. Some crystallographers consider anisotropic temperature factors as a sort of 'garbage bin' for measurement errors or, in other words, as a number of parameters that allow a nicely low disagreement factor between observed and calculated X-ray reflection intensities. There is some truth in this point of view, although it does not apply to data sets that were very carefully obtained and corrected for all possible errors.

To be on the safe side, we recommend the following procedure. Isotropic temperature factors are more reliable than anisotropic ones. Consider first these parameters of the heavy atoms (in an organic molecule C, N, O, etc., but not H), and calculate a scaling factor:

$$S_T = \frac{\sum_k U_{k,\mathrm{iso}}^{\mathrm{md}}}{\sum_k U_{k,\mathrm{iso}}^{\mathrm{exp}}} \tag{99}$$

After that one may decide to consider also a disagreement factor for the three diagonal anisotropic coefficients:

$$R_T = \frac{\sum_k \sum_{j=11,22,33} |U_{k,j}^{\mathrm{md}} - S_T U_{k,j}^{\mathrm{exp}}|}{\sum_k \sum_{j=11,22,33} |U_{k,j}^{\mathrm{md}}|}. \tag{100}$$

Experimental hydrogen positions are not always reliable, except for the few cases where the structure has been determined by neutron diffraction. In any case the X-ray bond distances, mostly between 0.6 and 1.0 Å, are too short since the average electron positions are detected rather than the nuclear positions. It is better to move the hydrogen coordinates along the observed direction of the bond in such a way that C–H bonds of about 1.1 Å and O–H bonds of about 1.0 Å are obtained [80] (see also Chapter 3, Section 1.1).

In a recent study [81] we performed molecular dynamics calculations on seven monosaccharides. Simulated and experimental results are compared in Table 4.1. In a few cases two experimental studies of the same substance were available, and a similar comparison could be made. Then the geometrical results agreed, of

Table 4.1 Comparison of observed and calculated crystal properties [81]

Δa	ΔV	Δr	$\Delta\theta$	$\Delta\phi$	S_T	R_T
3.3%	3.2%	0.24 Å	1.9°	3.7°	1.3	0.2

Average absolute differences between simulated and observed cell axes (a), cell volumes (V), atom coordinates (r), bond angles (θ), dihedral angles (ϕ), and the thermal indices S_T and R_T. Hydrogen atoms were excluded from the analysis.

course, much better, but the indices S_T and R_T were of comparable order of magnitude as in Table 4.1. So in a simulation of a known crystal structure only large deviations from the ideal values $S_T = 1$ and $R_T = 0$ should be a cause for concern: they may indicate that the crystal symmetry has been destroyed during the simulation.

The main purpose of our study was to obtain a better force field for carbohydrates, but it turned out that the simulations can suggest disorder not easily observable by X-ray diffraction and can even indicate experimental errors: *structure validation*. Such disorder was seen in the form of occasional transitions of dihedral angles between various local minima. In β-D-glucose and β-D-mannose some hydroxymethyl groups occasionally occupied deviating conformations. One would be inclined to conclude that the force field is inadequate and that torsional barriers should be increased, but inspection of average geometries and temperature factors did not reveal excessive deviations from experimental results. So an alternative explanation might be that such disorder is really possible, and not easily detected by X-ray analysis unless the effect is quite important.

In two cases, the molecular dynamics simulations resulted in hydrogen atom positions that were significantly different from the experimental ones. One case was α-D-galactose. Interestingly, two experimental structures had been published that differed in the positions of the hydroxyl hydrogen atoms, and therefore in the entire hydrogen-bonding scheme. Starting from one of them [82] the crystal structure remained unchanged, but starting from the other one [83] an entirely new hydrogen-bond scheme was obtained. The three schemes are shown in Figure 4.1. To settle the matter we prepared new crystals and redetermined the structure at liquid-nitrogen temperature, and found that the new structure predicted by molecular-dynamics simulation was indeed the correct one. In fairness, it should be added that Jeffrey and Shiono [84] had reached the same conclusion by a careful analysis of the oxygen positions and possible hydrogen-bonding schemes.

In the second case, α-D-glucose, our redetermination of the structure revealed no differences from the earlier one, but it did confirm the possibility of disorder in the hydrogen-bond network as suggested by the molecular-dynamics calculations.

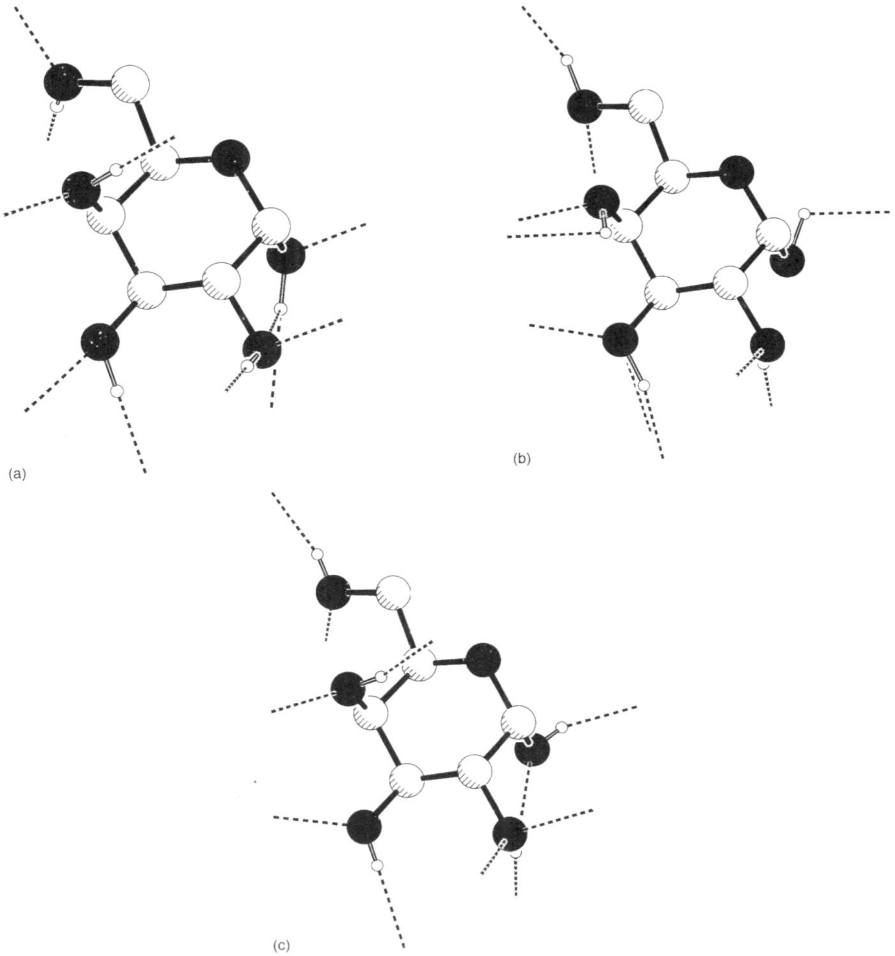

Figure 4.1 Three proposed hydrogen-bonding schemes for α-D-galactose; (a) Sheldrick [82], (b) Ohanessian and Gillier-Pandraud [83], (c) our MD simulations, in agreement with Jeffrey and Shiono [84]

4.2 Modeling of Crystal Structures: Cellulose

Cellulose, a polymer of $\beta(1 \rightarrow 4)$ linked glucose residues, is the most abundant macromolecule on earth. It has a strong tendency to form highly crystalline fibres. X-ray diffraction studies have been carried out on these fibres since about 1920. These studies have revealed that there are two major polymorphs: cellulose I, the native form, and cellulose II, which is invariably obtained after regeneration. Both crystal modifications have a monoclinic unit cell and space group $P2_1$ with cellulose molecules on twofold screw axes; there are two independent glucose residues (origin and centre) in the unit cell.

X-ray diffraction data usually do not go beyond a resolution of about 2.5 Å. This means that typically about 50 reflections are obtained, from which some 80 structural parameters should be extracted. This is clearly impossible unless additional information is used. Till now, the proposed models have only had limited structural detail. With state-of-the-art molecular modeling techniques, more reliable results can be obtained. In principle the problem can be approached in two ways.

One such method is used in the refinement of protein structures with limited-resolution X-ray data. Here the target function in the structure refinement consists of both experimental structure factor data and a potential-energy function, combined with appropriate weights. Usually, alternate cycles of MD simulations and structure refinements are carried out. An alternative approach is to perform MD simulations (or energy minimizations) on separate models that seem compatible with the diffraction data. For instance, from a few very strong reflection intensities it could be determined that the two cellulose chains are oriented in such a way that they run parallel or antiparallel and that they have a relative stagger of $\pm\frac{1}{4}\mathbf{c}$ (fibre axis). Thus only a limited number of models remain. The likelihood of the models and the degree of compatibility with X-ray fibre data can then be assessed by MD simulations. This is the approach taken here. An extensive MD study on possible polymorphs of cellulose is described elsewhere [85]; here only two aspects will be considered.

4.2.1 Cellulose I

On the basis of many experimental data of diverse nature, it gradually became accepted that the two independent cellulose molecules are oriented parallel. Two different structural models for the Iβ structure (there is also an Iα subform that we do not discuss here) were obtained from X-ray diffraction data. Gardner and Blackwell [86] preferred the parallel-up model (p1) whereas Sarko et al. [87, 88] preferred the parallel-down (p2) packing. These structures are only different because of the fact that the monoclinic angle γ deviates from 90°,* leading to a different packing in these two cases. The unit cell dimensions we used were taken from Gardner and Blackwell [86].

Both the p1 and p2 models were simulated at constant pressure (NPT) with GROMOS. All hydroxymethyl groups (exocyclic CH_2OH) were placed in the *tg* conformation,† for which there is ample experimental evidence. From the

*It is common practice to take *c* as the unique axis for fibres.

†The symbol *tg* stands for a *trans* conformation of O5-C5-C6-O6 and a *gauche* conformation of C4-C5-C6-O6.

Table 4.2 Average cell dimensions in MD

Cellulose Iβ	exp start	a 8.17	b 7.86	c(Å) 10.38	γ(°) 97.0	final
p1	*tg*/*tg*	8.1	7.6	10.4	84	*tg*/*tg*
p2	*tg*/*tg*	8.1	7.6	10.4	96	*tg*/*tg*

Cellulose II	exp start	8.01	9.04	10.36	117.1	final
a2	*gt*/*tg*	8.1	9.0	10.3	120	*gt*/*gt*
	gt/*gt*	8.1	8.9	10.3	119	*gt*/*gt*

simulations it is clear that p1 does not present a stable structure. The monoclinic angle γ (see Table 4.2) changes rapidly (within 10 ps) from 97° to 84°, with an associated shift of layers with respect to each other (see Figure 4.2). A view from the backside ($-\mathbf{c}$) shows that the structure has converged to the same monoclinic unit cell with the molecules placed in opposite c-direction. Close inspection of this structure reveals that it is just the p2 structure. Indeed, MD runs starting in the p2 structure are stable. Cell dimensions are reasonably close to the experimental ones; only the b-axis has shrunk by more than 0.2 Å. The *tg* conformation of the hydroxymethyl groups is well conserved. Clearly, on the basis of the MD simulations the p1 structure can be discarded.

4.2.2 Cellulose II

In this case most of the experimental data point at an antiparallel packing of the two independent cellulose molecules, as proposed by Kolpak and Blackwell [89] and Stipanovic and Sarko [90] (a2). In both studies the origin cellulose chains are found to possess a *gt* hydroxymethyl group conformation, whereas those of the centre chains have the *tg* conformation. Since other experimental data [91, 92] suggest that all hydroxymethyl groups are in the *gt* conformation, we here focus on this particular issue. The unit cell dimensions were taken from Kolpak and Blackwell [89].

The simulation starting in the a2 *gt*/*tg* structure gradually converged to the a2 *gt*/*gt* structure by conformational transitions of the hydroxymethyl groups in the centre chains (Table 4.2), although some disorder in this group remains, as can be seen by the anisotropic displacement factors from the MD simulation in Figure 4.3. The cell dimensions of the a2 *gt*/*gt* structure are close to experimental ones. Here the MD simulations have shown that indeed the *all-gt* conformation is preferred, and that this conformation should be used in refinement against X-ray data.

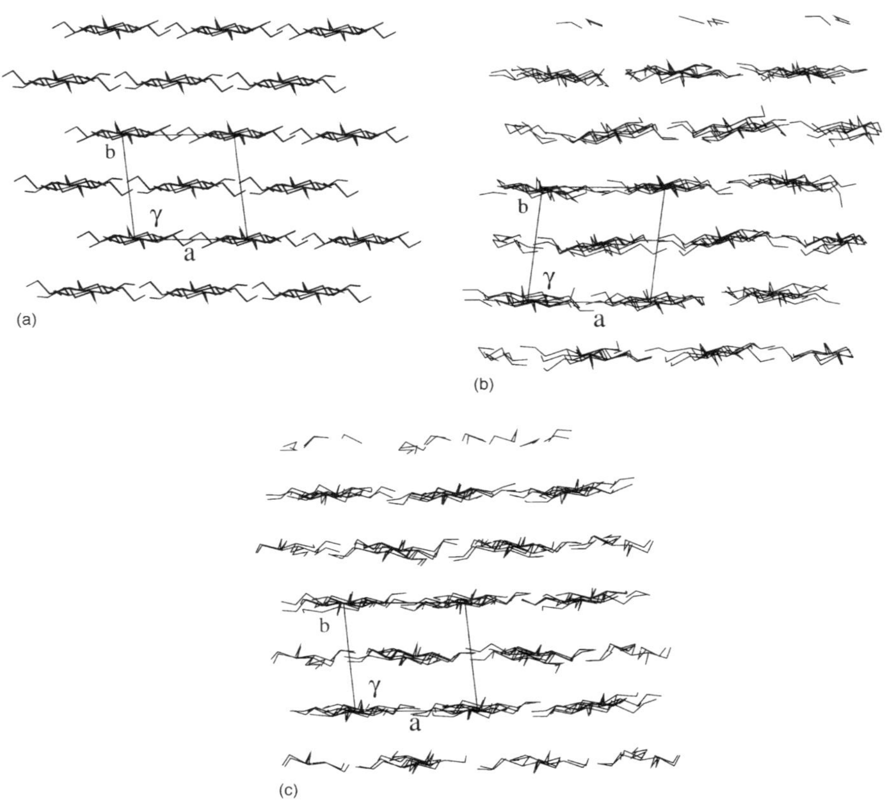

Figure 4.2 Projection down the **c**-axis of (a) the starting p1 structure, (b) a snapshot of the p1 structure after 100 ps that has converged to the p2 structure, and (c) a snapshot of the p2 structure after 100 ps

4.3 Prediction of Crystal Structures

In Section 4.1 we encountered one case where molecular-dynamics simulations, starting from an experimental structure, indicated that some hydrogen positions should be corrected. It must be noted that the structure changed from one local minimum to another one, so simple energy minimization could not have achieved this: in molecular dynamics, kinetic energy is available to overcome the energy barrier. Nevertheless, this can only work when the two structures are very similar and the differences between the local minima are small. Such a simulation can never find the transition from one polymorph to a completely different one, unless artificial high temperatures are used.

Here we are entering the subject of *ab initio crystal-structure prediction*. This

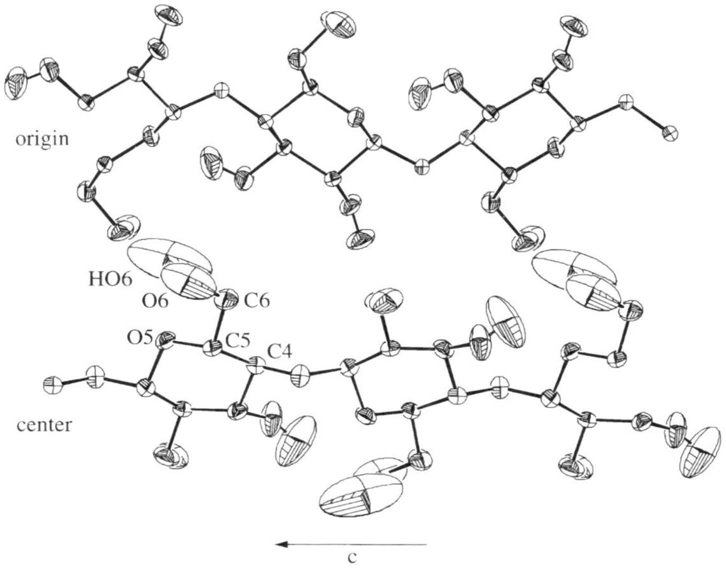

Figure 4.3 The two independent chains in the cellulose II (a2) structure. The thermal ellipsoids are drawn at a 50% probability level

is not *ab initio* in the quantum-chemical sense, because empirical force fields are used, but it means that no crystals have to be available. This is an exciting new field of research that was virtually nonexistent a few years ago [93]. Gavezzotti [94] was the first to report significant progress. He started by coupling pairs of molecules by one of a few symmetry operations, and selecting those with a promisingly low interaction energy for further consideration. A search for possible translations then leads directly to space groups with two molecules per cell, whereas space groups with four molecules per cell can be obtained after a second clustering of the molecule pairs. In favourable cases the crystal structures of hydrocarbons could be reproduced in this way. Independently, Gdanitz [95] developed a different approach where the space group is chosen in advance and the cell dimensions and the molecular positions are determined by Monte Carlo simulated annealing. He was able to reproduce the crystal structures of hexamethyl benzene and ethene; later on [96, 97] the method was extended with considerable success to molecules that also contain oxygen and nitrogen atoms. This work is now incorporated in a commercially available program package (see, for further details, Chapter 6).

Recently more papers on crystal-structure prediction have been appearing [98, 99, 100, 101]. Most work is done on structures that are constructed from rigid molecules, often with a pronounced shape that might limit the number of acceptable stacking patterns. Our paper on the prediction of hexapyranose structures [102] was the first where essentially flexible molecules were considered. In these molecules there are six internal degrees of freedom that play an essential role: the dihedral angle O—C—C—O which defines the conformation of the exocyclic CH_2OH group, and five C—C—O—H dihedral angles. Any fair attempt to predict such structures must consider these parameters as not known in advance.

One can distinguish two phases in a crystal-structure prediction, both quite difficult. The first phase is to make a list of structures that are not entirely unreasonable. This list was in older work rather short and obtained by hand, by comparing the molecule with related ones and guessing what kind of interactions could be possible and what kind of structures looked probable. A review of that type of work can be found in the book by Pertsin and Kitaigorodsky [4, p. 186]. In the recent papers cited above this subjective method has been replaced by more or less well-defined algorithms, always based on energies calculated by an empirical force field. It turns out that lists obtained in that way can contain hundreds of possible structures. Two strategies are used: random search and systematic search. The first method (MC or MD) is more flexible, but it can never be guaranteed to find the correct solution.

In the second phase the structures in the list are compared to find the most probable candidates for the experimental structure. This should be done by looking for the structure with the lowest free energy. It is not obvious that this is a sufficient criterion: the crystallization process need not produce the thermodynamically favoured structure, because in the beginning there must be a seed, a small molecular cluster, where kinetics is important and the energetics may be different. But, for the moment, this possibility is always disregarded. Furthermore, it is common practice to neglect the difference between energy and free energy. So, in fact, the calculated structures refer to 0 K and no phase transitions between polymorphs can be predicted. Calculated energies are not yet so accurate that one should start worrying about entropic terms – which, however, will eventually be necessary for predicting phase transitions.

All work reported so far indicates that the number of possible solutions is large, sometimes over a hundred within an energy range of 5 kcal/mole. This is equivalent to predicting a large tendency towards polymorphism. As an example we summarize the results of our work on hexapyranoses [102] in Table 4.3. In four cases out of six the force field used [81] was able to produce a good ranking for the experimental structure in the list of possible solutions, but in the two other cases it failed to do so.

In fact, this procedure is a new and very severe test for a force field: it should be able to select the observed crystal structure out of the list of possible ones. At the moment it is not clear whether or not kinetic effects might make this goal unachievable in principle, but a more modest demand would be to show that the

Table 4.3 Results of a crystal-structure prediction [102]

	R	E
α-D-galactose	2	0.2
α-D-glucose	2	0.2
α-D-talose	1	
β-D-allose	2	0.1
β-D-galactose	231	6.0
β-D-glucose	374	5.9

E is the energy (kcal mole^{-1}) of the experimental structure with respect to the structure of lowest energy, R is the corresponding ranking number of the experimental structure in the list of possible solutions.

energy of the experimental structure is not unreasonably higher than that of any other possible structure.

Although complete *ab initio* crystal-structure predictions are not yet reliable, the possibility of combining such information with X-ray powder data [98, 103] is definitely promising. Views on the prospects of crystal prediction vary from the optimistic statement of Karfunkel and Leusen that the molecular packing problem can be solved with the presently available computational methodologies [104] to Gavezzotti's temptation [105] to answer the title question of his paper 'Are crystal structures predictable?' with just the concise statement 'No'.

5. APPENDIX

5.1 Pressure Equations

In this section the pressure equations given in Section 2.1.1 will be derived. We do this for one unit cell with volume V; the results are easily generalized to a larger system. Consider the end planes R and L of the unit cell:

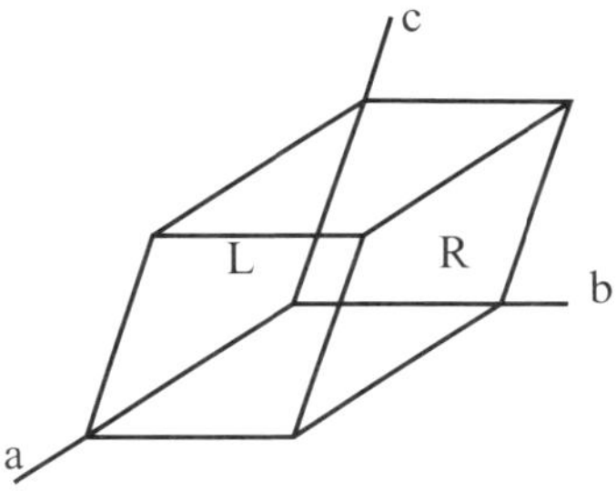

The outward vectors defining the surfaces are $\mathbf{N}^R = [\mathbf{c} \times \mathbf{a}] = \mathbf{b}^* V$ and $\mathbf{N}^L = -\mathbf{b}^* V$.

Let the axis **b** be changed by an infinitesimal vector d**b**. The surface R is then displaced by d**b**, so, following Eq. 26, the work performed on the system is:

$$w = -\mathbf{F}^R \cdot \mathrm{d}\mathbf{b} = -V \sum_\alpha \sum_\gamma P_{\alpha\gamma} b_\gamma^* \, \mathrm{d}b_\alpha = -V \sum_\alpha \sum_\gamma P_{\alpha\gamma} C_{2\gamma}^* \, \mathrm{d}C_{\alpha 2} \tag{101}$$

For changes in the three axes together:

$$w = -V \sum_g \sum_\alpha \sum_\gamma P_{\alpha\gamma} C_{g\gamma}^* \, \mathrm{d}C_{\alpha g} \tag{102}$$

The volume expansion is found by a similar reasoning:

$$\mathrm{d}V = V \sum_g \sum_\alpha C_{g\alpha}^* \, \mathrm{d}C_{\alpha g} \tag{103}$$

so in the isotropic case Eq. 102 reduces to the familiar expression $w = -P\,\mathrm{d}V$. Standard thermodynamics teaches us that Eq. 102 leads to:

$$\frac{\partial A}{\partial C_{\alpha g}} = -V \sum_\gamma P_{\alpha\gamma} C_{g\gamma}^* \tag{104}$$

which, by multiplication with $C_{\beta g}$ and summation over g, proves Eq. 27.

To develop this equation for a classical system, we need the partition function, Eq. 16. Here a problem is that the integration limits depend on the cell parameters. Following McQuarrie [11, p. 260], we change the variables **r** into the fractional coordinates **s** which are simply integrated from 0 to 1. The configurational part of the free energy is then:

$$A^{\mathrm{conf}} = -kT \ln \int \mathrm{e}^{-U(\mathbf{s},\mathbf{C})/kT} V^N \, \mathrm{d}\mathbf{s}_1 \ldots \mathrm{d}\mathbf{s}_N \tag{105}$$

from which we find by Eq. 27:

$$P_{\alpha\beta} V = \sum_g C_{\beta g} \left(\frac{NkT}{V} \frac{\partial V}{\partial C_{\alpha g}} - \left\langle \frac{\partial U(\mathbf{s},\mathbf{C})}{\partial C_{\alpha g}} \right\rangle \right) \tag{106}$$

which is identical to Eq. 28: the first term reduces to $NkT\delta_{\alpha\beta}$, the second one is identical to the virial expression given in Eq. 30.

Alternatively, we can apply the transformation to fractional coordinates not to every atom, but only to a certain vector in a molecule: an atomic position or the molecular centre of mass. For the other coordinates we take intramolecular distances, which are relatively small. After transformation to these variables Eq. 32 can be derived in the same way as above.

5.2 Strain and Stress

In the previous section we considered the pressure exerted by the crystal on its surroundings. A different view on the same subject is the effect of an applied external pressure, usually called the *stress* **S**; obviously, $\mathbf{S} = -\mathbf{P}$ (although this sign convention is not universally followed, see e.g. Ref. [71]). The resulting change in the equilibrium atomic coordinates can be formulated with aid of Eq. 9:

$$\Delta \mathbf{r}_{lk}^{\mathrm{e}} = \mathbf{C} \Delta \mathbf{s}_{lk}^{\mathrm{e}} + \Delta \mathbf{C} \mathbf{s}_{lk}^{\mathrm{e}} = \Delta \mathbf{r}_k^{\mathrm{e}} + \mathbf{u} \mathbf{r}_{lk}^{\mathrm{e}} \tag{107}$$

where $\mathbf{u} = \Delta\mathbf{C}\mathbf{C}^{-1}$. The first term is the internal deformation within a unit cell, the second term represents the deformation of the entire crystal lattice.

It is useful to transform the nine components of **u** into the new variables **e** and $\boldsymbol{\omega}$:

$$e_1 = e_{xx} = u_{xx}$$

$$e_2 = e_{yy} = u_{yy}$$

$$e_3 = e_{zz} = u_{zz}$$

$$e_4 = e_{yz} = u_{yz} + u_{zy}$$

$$e_5 = e_{zx} = u_{zx} + u_{xz}$$

$$e_6 = e_{xy} = u_{xy} + u_{yx}$$

$$\omega_{\alpha\beta} = u_{\alpha\beta} - u_{\beta\alpha} \tag{108}$$

Obviously $\boldsymbol{\omega}$ is antisymmetric, so for infinitesimal displacements $\boldsymbol{\omega}\mathbf{r}$ represents a trivial rotation of the entire crystal. Therefore, it is sufficient to consider the six elements of the **e** vector: the *strain* components. Likewise, since $S_{\beta\alpha} = S_{\alpha\beta}$ the nine stress components can be reduced to six, following the Voigt renumbering scheme indicated in Eq. 108.

From Eq. 102 follows the work done by the external force:

$$w = V\sum_{\alpha}\sum_{\beta} S_{\alpha\beta}u_{\alpha\beta} = V\sum_{i=1}^{6} S_i e_i \tag{109}$$

so that $VS_{\alpha\beta} = \partial A/\partial u_{\alpha\beta}$ or $VS_i = \partial A/\partial e_i$.

This is not the place for a further development of elasticity theory. We shall only define the *elastic constants* c_{ij} which occur in the generalized version of Hooke's law:

$$S_i = \sum_j c_{ij}e_j$$

$$\Delta U = \tfrac{1}{2}\sum_i\sum_j e_i c_{ij} e_j \tag{110}$$

with $c_{ij} = \partial^2 A/\partial e_i \partial e_j = V\partial S_i/\partial e_j$. Explicit expressions for the elastic constants can be developed from there, with aid of the following derivatives [71]:

$$\frac{\partial C_{\alpha g}}{\partial e_{\beta\gamma}} = \tfrac{1}{2}(\delta_{\beta\alpha}C_{\gamma g} + \delta_{\gamma\alpha}C_{\beta g}) \tag{111}$$

Interested readers should consult Kittel [13] and Landolt-Börnstein [10] for an introduction, Born and Huang [7] for a general treatment and Pertsin [4] for a development for molecular crystals with rigid molecules. Please note that there is hardly any standardization of notation in the literature. Kittel [13] uses ϵ rather than our **u**. Catti [106] interchanges this ϵ with **e**, introduces a factor $\frac{1}{2}$ in the lower four equations (108) and has $\boldsymbol{\tau}$ for the stress. Born and Huang [7] use **s** rather than our **e**.

6. REFERENCES

1. J. D. Dunitz, *X-ray Analysis and the Structure of Organic Molecules*, Cornell University Press, Ithaca, NY, 1979.
2. C. Giacovazzo, H. L. Monaco, D. Viterbo, F. Scordari, G. Gilli, G. Zanotti and M. Catti, *Fundamentals of Crystallography*, Oxford University Press, Oxford, 1992.
3. T. Hahn (editor), *International Tables for X-Ray Crystallography*, Volume A, Reidel, Dordrecht, 1983.
4. A. J. Pertsin and A. I. Kitaigorodsky, *The Atom–Atom Potential Method*, Springer, Berlin, 1986.
5. B. T. M. Willis and A. W. Pryor, *Thermal Vibrations in Crystallography*, Cambridge University Press, Cambridge, 1975.
6. A. I. Kitaigorodsky, *Molecular Crystals and Molecules*, Academic Press, New York, 1973.
7. M. Born and K. Huang, *Dynamical Theory of Crystal Lattices*, Oxford University Press, Oxford, 1966.
8. F. H. Allen and O. Kennard, *Chemical Design Automation News*, **8**, 31 (1993).
9. J. S. Chickos, 'Heats of Sublimation', in *Molecular Structure and Energetics*, edited by J. F. Liebman and A. Greenberg, pp. 67–150, VCH, New York, 1987.
10. M. M. Choy, W. R. Cook, R. F. S. Harmon, H. Jaffe, J. Jerphagnon, S. K. Kurtz, S. T. Liu and D. F. Nelson, *Elastic, Piezoelectric, Pyroelectric, Piezooptic, Electrooptic Constants, and Nonlinear Dielectric Susceptibilities of Crystals*, Landolt-Börnstein, New Series, Volume III-11, Springer, Berlin, 1979, edited by K.-H. Hellwege and A. M. Hellwege.
11. D. A. McQuarrie, *Statistical Mechanics*, Harper and Row, New York, 1976.
12. J. J. Erpenbeck and W. W. Wood, 'Molecular Dynamics Techniques for Hard-core Systems', in *Statistical Mechanics Part B*, edited by B. J. Berne, pp. 8–17, Plenum, New York, 1977.
13. C. Kittel, *Introduction to Solid State Physics*, John Wiley and Sons, New York, 1966.
14. B. Donovan and J. F. Angress, *Lattice Vibrations*, Chapman and Hall, London, 1971.
15. W. H. Press, B. P. Flannery, S. A. Teukolsky and W. T. Vetterling, *Numerical Recipes, the Art of Scientific Computing*, Cambridge University Press, Cambridge, 1989.
16. T. Schlick, 'Optimization methods in computational chemistry', in *Reviews in Computational Chemistry*, Volume III, edited by K. B. Lipkowitz and D. B. Boyd, pp. 1–71, VCH, 1992.
17. U. Burkert and N. L. Allinger, *Molecular Mechanics*, American Chemical Society, Washington, 1982.
18. W. F. van Gunsteren and H. J. C. Berendsen, *Angew. Chemie Int. Ed. Engl.*, **29**, 992 (1990).
19. M. P. Allen and D. J. Tildesley, *Computer Simulation of Liquids*, Oxford Science Publishers, Oxford, 1987.
20. W. F. van Gunsteren and H. J. C. Berendsen, *Mol. Phys.*, **34**, 1311 (1977).
21. J.-P. Ryckaert, G. Ciccotti and H. J. C. Berendsen, *J. Comput. Phys.*, **23**, 327 (1977).
22. P. H. Berens, D. H. J. Mackay, G. M. White and K. R. Wilson, *J. Chem. Phys.*, **79**, 2375 (1983).
23. B. P. van Eijck, *Molec. Simulat.*, **13**, 221 (1994).
24. T. P. Straatsma, H. J. C. Berendsen and A. J. Stam, *Mol. Phys.*, **57**, 89 (1986).
25. H. Flyvbjerg and H. G. Petersen, *J. Chem. Phys.*, **91**, 461 (1989).
26. H. J. C. Berendsen, J. P. M. Postma, W. F. van Gunsteren, A. DiNola and J. R. Haak, *J. Chem. Phys.*, **81**, 3684 (1984).
27. M. Karplus and J. N. Kushik, *Macromolecules*, **14**, 325 (1981).
28. A. DiNola, H. J. C. Berendsen and O. Edholm, *Macromolecules*, **17**, 2044 (1984).

29. W. F. van Gunsteren and P. K. Weiner (editors), *Computer Simulation of Biomolecular Systems*, Escom Science Publishers, Leiden, 1989.
30. M. J. Mitchell and J. A. McCammon, *J. Comput. Chem.*, **12**, 271 (1991).
31. L. A. Báez and P. Clancy, *Mol. Phys.*, **86**, 385 (1995).
32. L. Hua and G. S. Pawley, *Mol. Simulat.*, **7**, 89 (1991).
33. A. Boutin, J. M. Simon and A. H. Fuchs, *Mol. Phys.*, **81**, 1165 (1994).
34. N. L. Allinger, M. T. Tribble, M. A. Miller and D. H. Wertz, *J. Am. Chem. Soc.*, **93**, 1637 (1971).
35. N. L. Allinger, Y. H. Yuh and J. H. Lii, *J. Am. Chem. Soc.*, **111**, 8551 (1989).
36. O. Ermer and S. Lifson, *J. Am. Chem. Soc.*, **95**, 4121 (1973).
37. S. R. Niketic and K. Rasmussen, *The Consistent Force Field*, Springer, New York, 1977.
38. S. Lifson and A. Warshel, *J. Chem. Phys.*, **49**, 5116 (1968).
39. G. N. Ramachandran, C. Ramakrishnan and V. Sasisekharan, *J. Mol. Biol.*, **7**, 95 (1963).
40. G. Nemethy and H. A. Scheraga, *Biopolymers*, **3**, 155 (1965).
41. B. R. Brooks, R. E. Bruccoleri, B. D. Olafson, D. J. States, S. Swaminathan and M. Karplus, *J. Comp. Chem.*, **4**, 187 (1983).
42. S. J. Weiner, P. A. Kollmann, D. A. Case, U. Chandra Singh, C. Ghio, G. Alagona, S. Profeta and P. Weiner, *J. Am. Chem. Soc.*, **106**, 765 (1984).
43. W. F. van Gunsteren and H. J. C. Berendsen, GROMOS, Groningen Molecular Simulation Package, University of Groningen, 1987.
44. F. A. Momany, R. F. McGuire, A. W. Burgess and H. A. Scheraga, *J. Phys. Chem.*, **79**, 2361 (1975).
45. A. W. B. L. G. Dunfield and H. A. Scheraga, *J. Phys. Chem.*, **82**, 2609 (1978).
46. A. D. French, D. P. Miller and A. Aabloo, *Int. J. Biol. Macromol.*, **15**, 30 (1993).
47. A. T. Hagler, E. Huler and S. Lifson, *J. Am. Chem. Soc.*, **96**, 5319 (1974).
48. U. D. J. Maple and A. T. Hagler, *Proc. Nat. Acad. Sci.*, **85**, 5350 (1988).
49. M. Waldman and A. T. Hagler, *J. Comput. Chem.*, **14**, 1077 (1993).
50. M. L. C. E. Kouwijzer and P. D. J. Grootenhuis, *J. Phys. Chem.*, **99**, 13426 (1995).
51. U. C. Singh and P. A. Kollman, *J. Comp. Chem.*, **5**, 129 (1984).
52. A. Warshel and M. Levitt, *J. Mol. Biol.*, **103**, 227 (1976).
53. L. M. J. Kroon-Batenburg and J. A. Kanters, *J. Mol. Struct.* (THEOCHEM) **105**, 417 (1983).
54. A. Vedani and D. A. Huhta, *J. Am. Chem. Soc.*, **112**, 4759 (1990).
55. T. A. Halgren, *Curr. Opinion in Struct. Biology*, **5**, 205 (1995).
56. J. R. Maple, M. J. Hwang, T. P. Stockfisch, U. Dinur, M. Waldman, C. S. Ewig and A. T. Hagler, *J. Comp. Chem.*, **15**, 162 (1994).
57. M. J. Hwang, T. P. Stockfisch and A. T. Hagler, *J. Am. Chem. Soc.*, **116**, 2515 (1994).
58. T. A. Halgren, *J. Comp. Chem.*, **17**, 490 (1996).
59. W. D. Cornell, P. Cieplak, C. I. Bayly, I. R. Gould, K. M. Merz Jr., D. M. Ferguson, D. C. Spellmeyer, T. Fox, J. W. Caldwell and P. A. Kollman, *J. Am. Chem. Soc.*, **117**, 5179 (1995).
60. A. D. MacKerell, Jr., J. Wiórkiewicz-Kuczera and M. Karplus, *J. Am. Chem. Soc.*, **117**, 11946 (1995).
61. R. Destro and A. Gavezzotti, 'Intermolecular Energies and Packing Modes in Organic Crystals', in *Structure and Properties of Molecular Crystals*, edited by M. Pierrot, pp. 161–210, Elsevier, Amsterdam, 1990.
62. D. E. Williams, *J. Chem. Phys.*, **45**, 3770 (1966).
63. H. Bekker and P. Ahlström, *Mol. Simulat.*, **13**, 367 (1994).
64. T. Tominaga and S. Yip, *J. Chem. Phys.*, **100**, 3747 (1994).
65. D. E. Williams, *Acta Cryst.*, **A28**, 629 (1972).

66. J. L. Derissen, P. H. Smit and J. Voogd, *J. Phys. Chem.*, **81**, 1474 (1977).
67. B. R. A Nijboer and F. W. de Wette, *Physica*, **23**, 309 (1957).
68. D. E. Williams, *Acta Cryst.*, **A27**, 452 (1971).
69. M. W. Deem, J. M. Newsam and S. K. Sinha, *J. Phys. Chem.*, **94**, 8356 (1990).
70. M. L. C. E. Kouwijzer, B. P. van Eijck, S. J. Kroes and J. Kroon, *J. Comput. Chem.*, **14**, 1281 (1993).
71. N. Karasawa and W. A. Goddard III, *J. Phys. Chem.*, **93**, 7320 (1989).
72. C. M. Gramaccioli and G. Filippini, *Acta Cryst.*, **A35**, 727 (1979).
73. B. A. Luty, I. G. Tironi and W. F. van Gunsteren, *J. Chem. Phys.*, **103**, 3014 (1995).
74. G. Ruocco and M. Sampoli, *Mol. Phys.*, **82**, 875 (1994).
75. P. Ahlström, A. Wallqvist, S. Engström and B. Jönsson, *Mol. Phys.*, **68**, 563 (1989).
76. B. J. Palmer, *Chem. Phys.*, **184**, 163 (1994).
77. J. Applequist, J. R. Carl and K.-K. Fung, *J. Am. Chem. Soc.*, **94**, 2952 (1972).
78. B. T. Thole, *Chem. Phys.*, **59**, 341 (1981).
79. D. N. Bernardo, Y. Ding, K. Krogh-Jespersen and R. M. Levy, *J. Phys. Chem.*, **98**, 4180 (1994).
80. G. A. Jeffrey and L. Lewis, *Carbohydr. Res.*, **60**, 179 (1978).
81. M. L. C. E. Kouwijzer, B. P. van Eijck, H. Kooijman and J. Kroon, *Acta Cryst.*, **B51**, 209 (1995).
82. B. Sheldrick, *Acta Cryst.*, **B32**, 1016 (1976).
83. J. Ohanessian and H. Gillier-Pandraud, *Acta Cryst.*, **B32**, 2810 (1976).
84. G. A. Jeffrey and R. Shiono, *Acta Cryst.*, **B33**, 2700 (1977).
85. L. M. J. Kroon-Batenburg, B. Bouma and J. Kroon, *Macromolecules*, **29**, 5695 (1996).
86. K. H. Gardner and J. Blackwell, *Biopolymers*, **13**, 1975 (1974).
87. C. Woodcock and A. Sarko, *Macromolecules*, **13**, 1183 (1980).
88. A. Sarko and R. Muggli, *Macromolecules*, **7**, 486 (1974).
89. F. J. Kolpak and J. Blackwell, *Macromolecules*, **9**, 273 (1976).
90. A. J. Stipanovic and A. Sarko, *Macromolecules*, **9**, 851 (1976).
91. F. Horii, A. Hirai and R. Kitamaru, *Polym. Bulletin*, **10**, 357 (1983).
92. L. M. J. Kroon-Batenburg, J. Kroon and M. G. Northolt, *Polym. Commun.*, **27**, 290 (1986).
93. J. Maddox, *Nature*, **335**, 201 (1988).
94. A. Gavezzotti, *J. Am. Chem. Soc.*, **113**, 4622 (1991).
95. R. J. Gdanitz, *Chem. Phys. Lett.*, **190**, 391 (1992).
96. H. R. Karfunkel and R. J. Gdanitz, *J. Comput. Chem.*, **13**, 1171 (1992).
97. H. R. Karfunkel, F. J. J. Leusen and R. J. Gdanitz, *J. Comput.-Aided Mat. Des.*, **1**, 177 (1993).
98. J. R. Holden, Z. Du and H. L. Ammon, *J. Comput. Chem.*, **14**, 422 (1993).
99. T. Shoda, K. Yamahara, K. Okazaki and D. E. Williams, *J. Molec. Struct.*, **333**, 267 (1995).
100. K. D. Gibson and H. A. Scheraga, *J. Phys. Chem.*, **99**, 3765 (1995).
101. M. U. Schmidt, *Kristallstrukturberechnungen metallorganischer Molekülverbindungen*, PhD thesis, Technische Hochschule Aachen, Aachen, 1995.
102. B. P. van Eijck, W. T. M. Mooij and J. Kroon, *Acta Cryst.*, **B51**, 99 (1995).
103. K. D. M. Harris, M. Tremayne, P. Lightfoot and P. G. Bruce, *J. Am. Chem. Soc.*, **116**, 3543 (1994).
104. H. R. Karfunkel and F. J. J. Leusen, *Speedup*, **6**, 43 (1992).
105. A. Gavezzotti, *Acc. Chem. Res.*, **27**, 309 (1994).
106. M. Catti, 'Physical Properties of Crystals', in *Fundamentals of Crystallography*, edited by C. Giacovazzo, pp. 599–643, Oxford University Press, Oxford, 1992.

Chapter 5

Nucleation and Phase Transitions in Molecular Clusters: Molecular Dynamics Simulation and Experiment

LAWRENCE S. BARTELL
Department of Chemistry, University of Michigan

1. INTRODUCTION

Phase changes are among the most familiar transformations of matter known. Long before the flowering of science, man observed the transitions of freezing, melting, evaporation, and condensation, and made use of the phenomena even if there was no scientific conception of the meaning of phase. Today we are inclined to associate the phase behavior of matter with thermodynamics. To the ancients, however, what actually happened was more important than what was supposed to happen if things proceeded at equilibrium. Indeed, they devised some truly remarkable materials such as Damascus steel [1], which rivals the best of modern steels, by empirically discovered kinetic, rather than thermodynamic, control of products. It turns out that the subjects of this chapter, molecular clusters, may also attain thermodynamically unfavorable structures through the agency of kinetics. Of greatest concern in this chapter will be phase changes taking place spontaneously when a system is cooled. Thermodynamic equilibrium is much easier to maintain when clusters are heated than when they are cooled, for reasons to be discussed below. Even though a phase generated

Theoretical Aspects and Computer Modeling Edited by A. Gavezzotti

upon cooling is not always that with the greatest thermodynamic stability, thermodynamic considerations nevertheless enter prominently into our present notions about the kinetics of phase changes.

The thermodynamic treatment of phase transitions evolved rapidly after Joule's successful challenge of the caloric theory [2]. It was essentially brought to completion by the prescient developments of J. Willard Gibbs over a century ago [3]. On the other hand, the dynamics of phase transitions is a much younger science, and much less developed even today. This, despite the fact that it was Gibbs, himself, who contributed the pivotal insight into the problem by pointing out the free-energy barrier to the formation of a new phase in the body of an old phase [3]. Unfortunately, his perceptive observation went almost unnoticed for a half-century until Farkas [4], Volmer [5], and Becker and Döring [6], put considerations of the thermodynamic cost of forming the required nuclei of the new phase together with Einstein's formulation [7] of the probability of generating such nuclei via fluctuations, and devised a capillary theory for the rate of condensation of supercooled vapors. This capillary theory was generalized by Turnbull and Fisher [8], Buckle [9], and others to treat transitions in condensed matter. It is the latter types of phase change that are the main concern of this chapter. The capillary treatment is now known as the 'classical theory' of homogeneous nucleation. It has been found to describe qualitative features of nucleation excellently but lacks quantitative rigor, for reasons to be outlined below. One reason for the underdeveloped state of nucleation theory is the lack of a broad base of reliable experimental information to motivate and guide theorists to create a better, generally applicable theory. The trouble is that homogeneous nucleation is seldom responsible for phase changes encountered under ordinary circumstances. Extreme measures have to be taken if nucleation is to be accepted as genuinely homogeneous. Almost all phase changes seen in supercooled systems are heterogeneously initiated, catalyzed by some fortuitous contaminant [10].

It is apparent that the formulation of a satisfactory theory of homogeneous nucleation would be a worthy intellectual goal in an academic environment. But why would it be of any practical importance to develop such a theory if homogeneous nucleation is a rare phenomenon? For one thing the alternative, heterogeneous nucleation, which *is* an everyday event, is commonly treated as a development of homogeneous nucleation [10]; to the extent that homogeneous nucleation is poorly treated, ideas about heterogeneous nucleation must be correspondingly flawed. For another thing, it seems increasingly probable that significant atmospheric processes do involve homogeneous nucleation [11]. Even though many processes in science and technology depend upon nucleation, a recent editorial comment in the journal *Science* [12] lamented the fact that not even the simple process of freezing is understood on a molecular basis. Deficient though currently applied versions of nucleation theory may be, they do make it possible to make semiquantitative predictions about the behavior of systems subjected to greatly increased supercoolings. As will be shown, such predictions have proven to be helpful [13]. Analyses of nucleation rates based on nucleation theory also provide information of semiquantative accuracy about interfacial properties that have been extremely difficult to obtain by any other approach. Inasmuch as surface science is one of the most important in commerce, and one of the fastest growing in scientific laboratories, improvements in nucleation theory can be expected to be beneficial to many disciplines.

As mentioned above, one of the major obstacles to the development of nucleation theory is the difficulty in completing the conceptual framework of such a complex process without having an adequate body of empirical information available to guide the formulation. Despite a considerable effort by experimentalists, a large body of reliable information does not yet exist. It has proven to be extremely difficult to carry out experiments that are precise and unequivocally of homogeneous nucleation. Submicroscopic heterophase contaminants may, and all too often do, catalyze the formation of the new phase and spoil the results. It is all but impossible to prepare systems pure enough to ensure that heterogeneous nucleation cannot take place. One chance mote can cause an entire system to transform. It was Turnbull [14–16] who led the way by devising a technique suitable for studying nucleation in condensed phases. To get around the problem posed by chance contaminants he subdivided his liquid samples into droplets so small and numerous that most of them could be expected to be free from interfering particles. To keep the droplets separated from each other he suspended them in an emulsion with a surfactant to stabilize the suspension. Others explored alternative methods to keep the droplets apart but the possibility that the materials in contact with the droplets might themselves induce heterogeneous nucleation was always a worry. Tests were devised to analyze whether the nucleation rates were proportional to the volume or to the area of the systems of droplets. Nevertheless, it was sometimes found that systems of droplets separated by a new material were found to nucleate at a lower rate than prior systems apparently passing the tests. This was confirmatory evidence that the earlier experiments had been subject to heterogeneous nucleation (or catalysis by dissolved foreign ions).

Because of the troubles encountered in conventional nucleation experiments, it was felicitous (if accidental) that a new technique of promise was discovered. In any attempt to study the structures of liquids by electron diffraction [17] it is necessary to generate samples thin enough to transmit electrons cleanly, yet large enough to have bulk properties. Electrons are scattered approximately 10^8-fold more strongly by matter than are X-rays and neutrons, and hence cannot penetrate much further than a few hundred angstrom units without degrading the resultant diffraction pattern. It appeared that the only way to generate reproducibly thin and uniform samples would be to condense vapor into large molecular clusters in the chilling flow of a supersonic expansion. Such a technique had already been developed in several laboratories. Moreover, two research groups were probing the structures of the clusters so formed by electron diffraction [18]. Although these groups of physicists and gasdynamicists reported only solid clusters, not the liquid clusters we sought, we supposed that by resorting to more complex systems than those attractive to physicists we could generate clusters that were liquid. This not only proved to be true but we found, in addition, that the liquids could be studied while they were highly supercooled, a circumstance not possible in conventional X-ray and neutron diffraction investigations. An advantage of supercooling was the sharpening of the pair-correlation functions of the liquids, rendering their diffraction patterns more amenable to analyses of details about the interaction potential-energy functions characterizing the liquids.

Although this technique was quite successful, it led to observations that were even more interesting than the structures of liquids. Molecules of a given substance were sometimes seen to aggregate in the condensate in one of two, or occasionally, any of three different structures depending upon expansion conditions of the vapor [18, 19]. How could the way the vapor was passed through the throat of a supersonic nozzle control the way the condensing molecules decided to pack? It was quite clear that the critical nuclei first formed by condensation were too small and warm to be anything other than amorphous, liquid-like particles. As they grew, it was presumed that sooner or later they would attain a size sufficient for their packing arrangement to become identifiable with some bulk phase. It was ultimately established that the thermal history of the growing clusters did indeed depend upon the conditions of flow. What was seen when the clusters emerged from the nozzle depended upon the temperature of the clusters when they were large enough to possess a definite phase but this information was insufficient to dictate the phase. What was seen also depended upon the rate with which clusters could change phase during the cooling they experienced by thermal accommodation in the nozzle or as they evaporated in flight through the vacuum of the diffraction chamber.

It was eventually realized that, in several examples already examined, a transition *must* have taken place inside the supersonic nozzle [19]. When this was recognized, conditions were adjusted to induce the transitions to occur at a convenient location beyond the supersonic nozzle so that they could be monitored by electron diffraction. Varying the distance between the nozzle tip and the probing electron beam gave a time-of-flight measure of the temporal evolution of the system. Rates of both freezing and of certain solid-state transitions could be measured. Moreover, the freedom of the clusters from contact with any phase other than the rarefied neat vapor of the material and, sometimes, highly rarefied neon gas, assured that the nucleation was homogeneous.

Although this new technique made it possible to monitor the change in structure of clusters as they underwent transitions, the diffraction technique was blind to the minute structural fluctuations responsible for the nucleation. What was desirable was a technique that would reveal in molecular detail the processes leading to nucleation. Such a technique – a computational technique – already existed [20]. After experimental observations of a number of examples of nucleation occurring in supersonic jets, it was possible to apply nucleation theory to make rough projections of nucleation rates at deeper supercoolings. These predictions suggested which of the substances were likely to nucleate at deep supercooling during the million-fold briefer time-scales corresponding to molecular dynamics (MD) simulations. Attempts to observe nucleation events were then carried out on selected molecular clusters by MD simulations. They proved to be successful [13, 21–23].

Although these new cluster techniques afford promising alternatives to the conventional emulsion technique, they are not without their own problems. For one thing, clusters produced in supersonic flow and, to an even greater extent, those generated in the computer simulations, are so much more highly supercooled than the droplets investigated in previous techniques that the nucleation rates are higher by twenty to thirty or more orders of magnitude. It is reasonable to expect that such astronomically higher rates refer to the same mechanism? Have the results of such measurements anything to tell us about nucleation in ordinary transitions? We shall see that there is strong reason to believe that the answer is 'yes' to both questions.

2. WHY STUDY CLUSTERS?

Nucleation in condensed phases can be studied in clusters both experimentally by probing supersonic jets [24–29] and computationally [21–23], by molecular dynamics techniques. Advantages and limitations of the cluster techniques will be discussed in some detail after the methods are outlined. Before describing the methods, however, it is helpful to relate why it is useful to carry out research on clusters as well as on enormously larger systems of the type investigated in the past. Quite apart from being able to study homogeneous nucleation in new regimes of supercooling, one of the main driving forces has been to develop techniques that assure that the nucleation is homogeneous. As discussed in the introduction, it has proven to be extremely difficult to be entirely certain that nucleation is homogeneous when applying the traditional techniques. This drawback, combined with the difficulty of carrying out accurate measurements with these techniques, made it attractive to explore alternative avenues to nucleation. Clusters provide systems for which it can be certain that nucleation is homogeneous. Moreover, in computational investigations, the molecular behavior during nucleation and growth of the new phase can be followed directly and in detail. Indeed, inferences about what the molecules are doing at the critical stage can be drawn not only by numerical analyses of the MD simulations, but can be assimilated in an intuitively more appealing way by watching moving pictures of the molecules as they rearrange. The movies are produced from the MD output [30].

Other reasons for studying clusters have to do with the fact that the technologies for their study existed at the outset of the investigations in nearly the same form as they are practiced today. Therefore, a long period of development was unnecessary once the utility of the techniques to research on nucleation was recognized, and satisfactory procedures for extracting much of the relevant data were already in place. The well-developed electron diffraction method for determining structures of gas-phase molecules required only comparatively minor changes for it to be applied to clusters produced in a supersonic jet, and the conversion to clusters was initiated long before the idea of nucleation research was conceived of [18, 31]. The computational method had also been developed many years before studies of nucleation had attracted attention, although the computational power required was not initially available [20]. When nucleation studies were first attempted by MD simulations they were carried out on atomic, not molecular, systems, and the systems were bulk, not clusters. To perform bulk computations while keeping computational demands within reasonable bounds, it is necessary to limit the number of molecules incorporated and to impose periodic boundary conditions. These conditions seriously interfere with results. Sometimes interactions across the boundary artificially enhance the rate and sometimes, unless the system is impractically large, they frustrate it, particularly if the transformation is from a higher to a lower symmetry. Clusters with free boundaries avoid such troubles although they introduce some uncertainty in the fraction of the cluster volume that can be considered to be effectively available for nucleation. Before discussing further advantages and limitations of clusters, it is helpful to describe the way the cluster techniques work. This is done in Sections 3 and 4.

3. EXPERIMENTAL METHOD

3.1 Supersonic Jet

Clusters are generated by condensation of vapor in the cooling flow in a supersonic expansion [32]. Some vapors do not condense to clusters unless they are seeded into a large excess of carrier gas able to remove the heat of condensation. Others, such as argon, ammonia, carbon dioxide, and water, readily condense to clusters even in the absence of a carrier gas. Any gas flowing from an orifice into a vacuum undergoes a supersonic expansion if its concentration is great enough for the molecular mean-free-paths to be short compared with the diameter of the jet. It is sometimes supposed that the cooling effect in nozzle flow is the Joule–Thomson effect, an effect depending upon the nonideality of the vapor. Even though nonideality may have an effect (usually minor) on the gasdynamics involved it should be understood that the cooling associated with the supersonic expansion is not that of the Joule–Thomson effect. Under the most common experimental conditions, friction in the flow is not important and the irreversible expansion into a vacuum can be treated *during the transitory period of the flow* as if it were a reversible adiabatic expansion. For such an isentropic process, the gain in entropy accompanying the volume increase is compensated by the decrease in thermal entropy. Put in a more intuitive way, as an element of gas expands, it does work on the gas in front of it, and it pays for the work by losing heat. Such considerations apply to ideal gases as well as to nonideal gases. In fact, the nearly ideal rare gases exhibit the greatest temperature drops in supersonic expansions because they have the lowest heat capacities to furnish the enthalpy lost in accelerating the gas downstream. Of course, after the gas strikes a wall, its molecules recover their random directions of flight and the gas very nearly recovers its original temperature.

Different laboratories produce supersonic jets by different techniques. The commonest way is simply for the gas to be introduced into an evacuated chamber through a small orifice. Nearly as common is for it to be introduced through a short tube. The tube tends to concentrate the gas into a somewhat narrower stream than an orifice does if the flow is in the viscous regime, but at higher pressures the effluent broadens to a distribution resembling that produced by a pinhole. Until the gas exits the tubular nozzle it does not reach supersonic speeds or cool. In several laboratories including the author's, the gas is usually expanded through a miniature Laval nozzle, whose inner diameter increases monotonically beyond the throat, so that the flow becomes supersonic inside the nozzle. The advantage is that the nozzle wall slows the gas expansion, forcing the molecules to undergo many more collisions for a given degree of cooling. This makes it possible for clusters to be formed under much milder expansion conditions and decreases the demand for fast vacuum pumps. The disadvantage is that some of the transitions of possible interest may occur inside of the nozzle itself, preventing their direct observation.

3.2 Detection of Phase Change

The method that has been used most frequently to date to monitor changes in clusters in studies of nucleation, is electron diffraction. A promising alternative method makes use of high-resolution coherent Raman spectroscopy [33]. Both techniques are able to identify the phases encountered and the relative concentrations of the phases. The spectroscopic technique gives a better direct measure of the temperature of liquid clusters than does electron diffraction, although results so far published appear somewhat to overestimate the temperatures of clusters because of the extra freedom of the numerous surface molecules in comparison with those in the bulk calibrant. The diffraction technique, however, has the advantage of providing a direct measure of the particle size. Particle diameters D

are inversely proportional to the breadths of the Debye-Scherrer peaks of the diffraction patterns of the crystalline clusters in accordance with the relation (for spherical clusters)

$$D = 0.553/W_\theta \tag{1}$$

where W_θ is the full width at half-maximum of the Bragg peaks expressed in terms of the conventional X-ray variable $(\sin\theta_B)/\lambda$ with θ_B representing the Bragg angle, not the scattering angle. The availability of this information is the reason electron diffraction has so far been employed more frequently in nucleation research. Particle size must be known before nucleation rates can be deduced. In the remainder of this chapter only the diffraction method will be discussed.

3.3 Measurement of Diffraction Pattern and Inference of Time

Figure 5.1 illustrates schematically how electron diffraction experiments are carried out. A cluster beam in the supersonic jet emerging from the nozzle is intersected by a 40 kV electron beam. Electrons diffracted by the jet are recorded on a photographic plate, photodiode array (PDA), or charge-coupled device (CCD). Successive patterns (Figure 5.2) are recorded at increasing distances and, hence, at increasing times of flight between the nozzle and electron beam. Times of flight can be calculated from measurements of cluster velocities which, in turn, are derived by chopping the supersonic jet and determining the time it takes clusters to arrive at a fast ionization gauge downstream. Under typical conditions

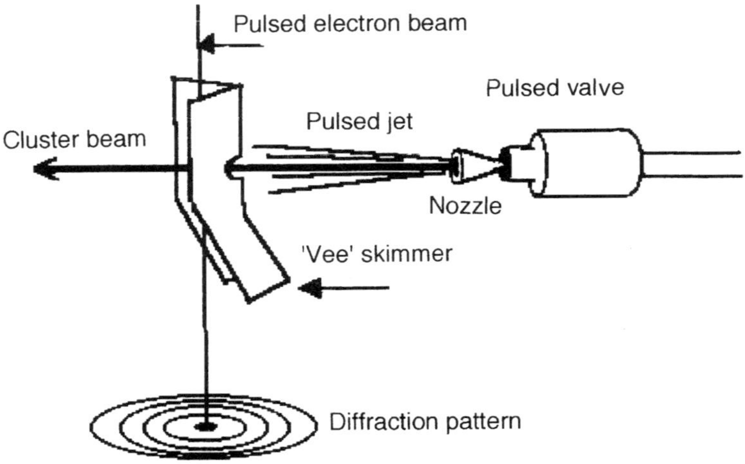

Figure 5.1 Schematic diagram of electron-diffraction monitoring of cluster beam

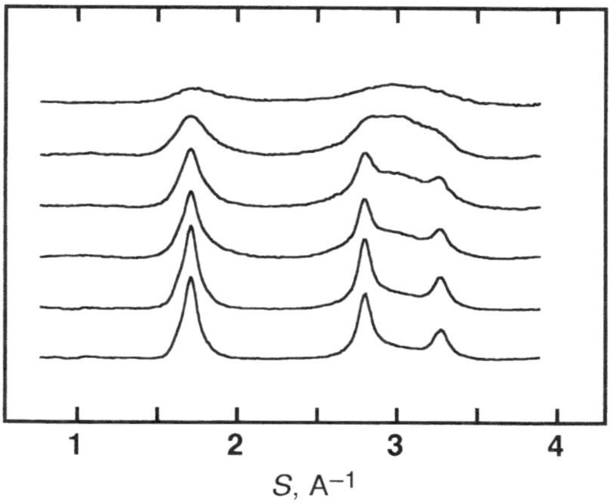

Figure 5.2 Electron-diffraction patterns of water clusters taken at 3.6 μs time-of-flight intervals, illustrating the transition from liquid to cubic ice

of breadth of electron beam and cluster velocity, the time resolution is of the order of a microsecond and it could be made more delicate if the need arose. Because the time for a transformation to be complete in a given cluster, once a critical nucleus has appeared, is short compared with the time for the whole ensemble of clusters to transform, a determination of the rate of conversion of the ensemble is a measure of the nucleation rate.

It is helpful to pulse the supersonic jet, and to probe the cluster pulses at carefully adjusted times by a synchronously pulsed electron beam. Such an arrangement ensures that the electrons sample the clusters cleanly before gas from the supersonic beam has had time to scatter from the walls of the apparatus and to build up a steady-state background. Background gas contributes an undesirable extraneous intensity and can cause shock waves that interfere with clusters. Between pulses, the vacuum is restored by fast pumps. Naturally, recording diffraction patterns with a CCD or PDA device is much more convenient than recording on a photographic plate because the direct recording enables the response to adjustments of experimental conditions to be seen almost at once [34]. Adjustments to optimize conditions when photographic recording is used are made blindly and are tedious. Recording electron intensities by direct impact of the electrons on CCD or PDA detectors is inadvisable. To avoid damage to the detectors and to amplify the number of quanta recorded, electrons strike a screen coated with an aluminized phosphor, and are detected by the emitted photons, perhaps a thousand for each electron impact. To maximize the number of photons reaching the detection device while preserving the spatial distribution of the diffraction pattern, photons are led to the CCD or PDA by a fiberoptic bundle.

3.4 Control of and Inference of Temperature

In cluster research the operator does not exercise control over temperature in the way as in conventional experiments on systems in contact with a heat bath. Instead, the operator is limited to the temperature range dictated by the properties of the evaporative ensemble of clusters. Under normal conditions of preparation, clusters growing in a supersaturated jet are kept warmer than the expanding gas surrounding them by the continually evolving heat of condensation – how much warmer depends upon the concentration of the carrier gas (if any). Once clusters of typical size have left the nozzle and have coasted a few centimeters into the vacuum of the diffraction chamber, they cool by evaporation to a temperature in the vicinity of the so-called evaporative cooling temperature, T_{evp} [35–37]. In doing so they may lose about 2% of their diameter. If the transition of interest occurs sufficiently rapidly somewhere in the range between the temperature of the warm new clusters and T_{evp}, there is an opportunity of seeing it. Only when the transition is close to T_{evp}, however, is the monitoring simple and convenient. It may be possible to cool clusters below T_{evp} by thermal accommodation in a high pressure of carrier gas. Despite the several papers suggesting formulas for it, and invoking it, the evaporative cooling temperature itself is not, of course, an invariant quantity. It depends upon the cluster size and lifetime before observation, and the size can be controlled by the conditions of expansion. Therefore, adjustments of experimental conditions can somewhat extend the number of systems that can be studied by the supersonic technique, but it is clear that the method is far from universally applicable.

Before discussing how the temperature is inferred in a typical experiment it is worthwhile to present several rough and ready methods for estimating T_{evp}. Such estimates give a quick idea of whether a given transition is likely to be accessible by the present technique. The idea behind the estimates is that a warm cluster with a comparatively high vapor pressure begins to evaporate rapidly, losing heat of vaporization as it does so. This loss of heat cools the cluster very quickly, considerably reducing its vapor pressure before a large proportion of its mass is sacrificed. The initial rate of cooling is so fast that the starting temperature is unimportant. What matters is that the temperature soon falls to a point where the volatility has become too low for appreciable evaporative loss on time scale of the experiment, and the rate of change of temperature has become small. A representative evolution of temperature with time is illustrated in Figure 5.3. Following the lead of Farges et al. [35] and Gspann [36], Klots [37] suggested that the temperature of an evaporative ensemble of clusters in the 100 Å range tends to be

$$T_{\mathrm{evp}} \approx 0.04 \Delta E_{\mathrm{vap}} / R \tag{2}$$

after some dozens of microseconds. Obviously, if such a rule applied fairly accurately to normal liquids, its Trouton basis would fail in the case of strongly

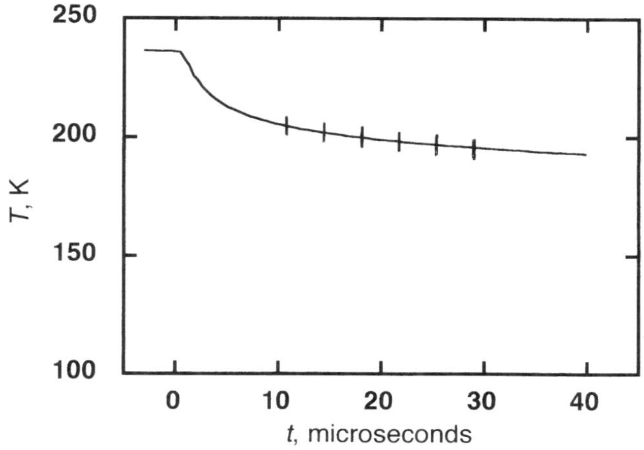

Figure 5.3 Temperature of liquid water clusters in their trajectories through space. The time origin corresponds to the emergence of clusters from the nozzle. Ticks on the curve indicate times at which the patterns in Figure 5.2 were recorded

associated liquids and solid clusters. An alternative rough rule of thumb was suggested to compensate for effects of association or state of aggregation, namely that T_{evp} is approximately the temperature at which the vapor pressure has fallen to about 0.4 Pa [24]. A more detailed treatment of evaporation rates shows that significant elements of the problem are left out of both of the above approximations [38]. Nevertheless, both serve as useful guides.

The method currently used at the University of Michigan to estimate the evolution of the temperature profile of the clusters during the course of their formation and growth, and their flight after they leave the nozzle is rather complex and will be presented here in bare outline. It gives results broadly consistent with the above estimates of T_{evp} but is believed to be more reliable. It also makes reasonable predictions of the sizes of clusters generated under various expansion conditions. The treatment of nucleation and cluster growth within the nozzle has been described in detail in the literature [38]; a comparable treatment after clusters leave the nozzle has not yet been published but is simpler, its principal part being just that of the rate of evaporation of small clusters into a rarefying gas. Inside the nozzle the seeded gas expands according to the well-known gasdynamic laws conserving energy, mass flow, and momentum (except for small frictional losses at the nozzle wall) until the flow becomes cool enough to begin to form critical, liquid-like nuclei. Nucleation continues as the existing nuclei grow in the supersaturated vapor and evolve heat into the surrounding gas by thermal accommodation. Ultimately, the warming of the gas shuts down further nucleation while the clusters already produced continue to grow until they exit the nozzle or exhaust the vapor. A dynamic process of exchange of energy and mass between the clusters and the expanding gas is taken into account as condensation, evaporation, and thermal accommodation take place. The simplifications usually found in treatments of cluster formation, namely that the cluster temperature and velocity are the same as those of the surrounding gas, are not made. Because the rate of extraction of heat from clusters is limited by the rate of gaskinetic collisions, the rate of deposition of heat by condensation outpaces thermal accommodation and keeps clusters appreciably warmer than the gas. At some point in the expansion a 'quitting surface' is reached where the momentum exchange between clusters and the rarefying gas is insufficient to continue to accelerate the clusters as the gas expands (and accelerates). This is taken to be the point in the flow at which the calculated gas velocity is the same as the experimentally measured cluster velocity, and from this point in calculations, clusters are allowed to lag behind the expanding gas.

The essential ideas of the procedure to determine cluster temperature, then, are that under normal operating conditions the clusters exit the nozzle at temperatures markedly warmer than the evaporative cooling temperature. Soon the carrier gas is so rarefied that the problem reduces to the relatively simple problem of cooling rate by evaporation into a vacuum. The thermal history of clusters outside the nozzle is computed by classical kinetic theory of evaporation and condensation. Because this treatment leads to exactly the same expression as that derived by Klots from quantum partition functions [39], we can have fair confidence in the results. An experimental check is available for clusters such as the hexafluorides which emerge from the nozzle as crystalline. Their temperatures derived from lattice constants are the same as their evaporative cooling temperatures to within the range of experimental error [40]. Occasionally, the carrier gas pressure is enormous in comparison with that of the vapor, and clusters leave the nozzle colder that T_{evp}. Under these conditions our accuracy in computing the temperature profile is substantially diminished.

3.5 Analysis of Data

Recorded electron diffraction patterns of clusters undergoing a phase change are reduced to a standard 'leveled' form, then analyzed to find the fraction to have transformed. From the rate of transformation is derived the nucleation rate which can be interpreted in terms of classical nucleation theory to derive the kinetic 'interfacial free energy' parameter characterizing the boundary between the old phase and the critical nucleus. It is this parameter, out of all proportion to the others, that dictates how nucleation rate varies with temperature. It is useful for estimating whether the nucleation rate at deeper supercooling is likely to be high enough for the transition to be simulated successfully on the much shorter timescale of molecular-dynamics simulations. Alternatively, if experimental data exist from conventional studies of nucleation in small droplets, it is worthwhile to learn whether the cluster studies and studies made on small droplets can be consistently related to each other via existing nucleation theory. In the following, the more important of the steps in analysis will be described briefly.

Leveled diffraction patterns of clusters of water [28] are reproduced in Figure 5.2 to illustrate the steps in processing the data. They were recorded at intervals, beginning when the clusters were liquid and ending when most of the clusters had frozen to cubic ice. Given a pattern recorded at time t, the fraction $F(t)$ of clusters to have transformed is determined from the optimum coefficients of a linear combination of reference patterns corresponding to liquid and solid. Once $F(t)$ is known, the nucleation rate J and interfacial free energy can be determined together, making use of nucleation theory in the curve-fitting of the $F(t)$ data. As will be shown subsequently, the nucleation rate $J(T)$ is calculated, according to the classical theory of homogeneous nucleation, by the expression

$$J(T) = A \exp(-\Delta G^*/kT) \tag{3}$$

where the prefactor A and the free-energy barrier ΔG^* to the formation of a critical nucleus depend upon the interfacial free-energy parameter σ (which is assumed to be unknown) and other thermodynamic quantities that can be estimated from existing information. It is not crucial to the determination of nucleation rate whether the theoretical expression is highly accurate or not. The point of invoking the expression is simply to be able to pass a plausible calculated curve through the somewhat scattered experimental $F(t)$ points (which points each correspond to slightly different temperatures because of the temperature drop of the clusters with time). The definition of $F(t)$ for a system of N_T identical clusters undergoing a change from phase 1 to phase 2, is

$$F(t) = N_2/N_T \tag{4}$$

while the definition of nucleation rate, or rate of production of critical nuclei per unit volume of untransformed material, is

$$J = (\mathrm{d}N^*/\mathrm{d}t)/V \tag{5}$$

or

$$J = (\mathrm{d}N_2/\mathrm{d}t)/(N_T - N_2)V_\mathrm{c} \tag{6}$$

by virtue of the fact that the growth time of the new phase in a nucleated cluster is short in comparison with the time for the whole ensemble to change, where V_c is the volume of a single cluster. In order to integrate Eq. (6), the expression (3) for $J(T)$ is transformed from a function of temperature to a function $J(t)$ of time using the thermal profile $T(t)$ of clusters (Figure 5.3) calculated as discussed in the foregoing. Then, from Eqs (4) and (6) it follows that

$$F(t) = 1 - \exp\left[-\int_0^t J(t)V_\mathrm{c}\,\mathrm{d}t\right]. \tag{7}$$

A least-squares fitting of the experimental data for $F(t_i)$ by the calculated expression through adjustments of parameter σ can then be performed, carrying out the integration in Eq. (7) numerically. Over the range of experimental data, then, the best estimate of the nucleation rate is just that implied by the integrand of Eq. (7), and the result is not particularly sensitive to the variant of the theoretical expression in Eq. (3). How accurately $J(T)$ extrapolates to much higher or lower temperatures does depend upon the variant used.

4. COMPUTER SIMULATION OF NUCLEATION

4.1 Molecular Dynamic Computations

Two types of numerical simulations are in widespread use in computations of the thermodynamic properties of systems of molecules interacting with each other at a given temperature or total energy. Although there are many variants of each, we will consider only the commonest. One is the Monte Carlo method in which, as it were, a mathematical stick stirs the system until its successive configurations of molecules conform to a Boltzmann distribution in potential energy. Although this method has produced valuable thermodynamic information about systems undergoing phase changes, it does not explicitly introduce the element of time. Therefore, it is not the method of choice in studies of rate processes.

The other method, the molecular dynamics (MD) method, directly follows the trajectories of molecules as a function of time. Although procedures to compute the time evolution of a quantum wave packet representing a cluster exist, they are computationally extravagant. Moreover, in the case of comparatively warm systems of molecules with molecular weights large in comparison with that of helium, quantum results have been shown to be indistinguishable from the results of classical molecular dynamics. Therefore we shall consider only the latter.

In the MD method as it is applied to clusters, one begins by arranging the molecules into some starting spatial configuration with an initial distribution of translational and rotational velocities approximating that for a Boltzmann distribution. The assembly of molecules then moves at constant energy in accordance with the laws of Newtonian mechanics. Molecules collide with each other again and again, redistributing their energies chaotically. If the initial configuration is not efficiently packed, molecules spontaneously fall into a more stable configuration, acquiring kinetic energy as they lose potential energy. The temperature of the aggregate is computed from the mean kinetic energy in accordance with the law of equipartition of energy. A typical run might begin by allowing molecules to equilibrate for a while in what amounts to a heat bath at the desired temperature, then freeing them for a few thousand time steps to carry out their trajectories at constant energy. In the heat bath kinetic energy is added or bled away to adjust the system to the desired temperature. Analyses are simpler if the initial angular velocity of the cluster as a whole is constrained to be zero. Conservation of angular momentum preserves this state. Because thermal kinetic energies of rotation of clusters (averaging $3kT/2$) are too low to have an appreciable effect on internal properties, the neglect of the centrifugal distortion poses no problems.

In order to perform MD simulations, then, it is necessary to find or construct a potential-energy function that expresses the intermolecular interactions realistically enough for the purposes of the problem to be solved. Not long ago, the forms of potential-energy functions adopted tended to be simple in order to make the demanding computations feasible. As computational speed increases, more complex functions are becoming affordable. In the systems of small molecules to be considered in this chapter, it is assumed to be adequate to treat the molecules as rigid bodies. If molecules were made flexible by giving them realistic intramolecular force fields, and if they were then allowed to undergo classical intramolecular motions as they collided classically with their neighbors, the results would be inferior to the results for rigid molecules. Vibrational excitation energies in such small molecules are typically enough larger than thermal energies at the temperatures of concern that molecules are more accurately regarded as rigid than as able to absorb or give up arbitrary amounts of energy in collisions.

The molecular clusters treated in MD runs in this chapter are bound by van der Waals forces and range from perhaps 100 to 500 molecules in size. Cold *atomic* van der Waals clusters of this size organize spontaneously into noncrystalline, polyicosahedral aggregates unrelated to the stable bulk forms [41]. By contrast, systems of *polyatomic molecules* of this size exist stably in the same structures as the bulk systems [13]. Although surface layers tend to be disordered, the interiors of the clusters investigated to date closely resemble the bulk in energy per molecule and in rotational diffusion as well as in structure. There is usually no ambiguity in identifying the phase of a cluster because the properties correspond so closely to those of bulk phases. Therefore, molecular clusters are not only convenient subjects for study but are also realistic models of bulk condensed matter. To be sure, there are pronounced size effects on transition temperatures, and transitions do not exhibit the sharp discontinuities characteristic of the bulk as the temperature changes. But these differences are expected, moderately well understood, and easily allowed for [42].

4.2 Analyses of Simulations

To find whether phase changes occurred while a cluster was being heated or cooled, one of the first steps is to plot $U(T)$, the mean configurational energy (potential energy) as a function of the temperature. In the typical plot presented in Figure 5.4 can be seen inflections corresponding to a solid-state transition (at the lower temperature) and melting [23]. For the comparatively small clusters

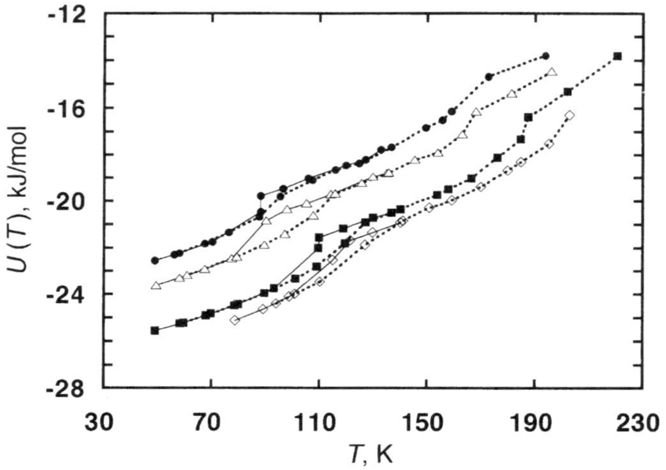

Figure 5.4 Configurational energies of $(SeF_6)_N$ clusters during runs of heating (dashed) and cooling (solid), illustrating the transitions between monoclinic and b.c.c. (at about 80–100 K) and melting (at about 170–200 K). Curves, from top to bottom, $N = 150$, 250, 350, and 500

amenable to MD simulations the transitions corresponding to first-order transitions in the bulk phase do not exhibit the sharp discontinuities in $U(T)$ characteristic of the bulk. For large clusters this inflection may be conspicuous but for the smaller clusters, it may be subtle. Procedures for identifying the phases involved, and for recognizing the transitions with greater delicacy than afforded by inspection of $U(T)$ are described below.

Many techniques are available to determine whether a cluster is liquid or crystalline and, if crystalline, which bulk phase it is related to, and the reader can no doubt suggest more than are mentioned in the following. For the first quick assessment a visual image of the structure is informative. With the aid of a proprietary program such as MACSPIN, projections of the molecular coordinates upon the screen of a monitor give an image of the cluster, and the cluster's orientation can be arbitrarily manipulated while watching the image. The difference between a randomly packed liquid and an ordered crystal is obvious at once, of course. An image of a 188-molecule cluster of *t*-butyl chloride undergoing a transition from its tetragonal phase to monoclinic [43] is presented in Figure 5.5.

To distinguish quickly between various crystalline phases it is particularly convenient to inspect the Pawley projection [44], especially if the molecules are highly symmetric. Vectors along all bonds directed towards an imaginary hemisphere high above a cluster are projected onto the hemisphere, then projected back down to a plane through the equator of the hemisphere. The 'dot plot' of

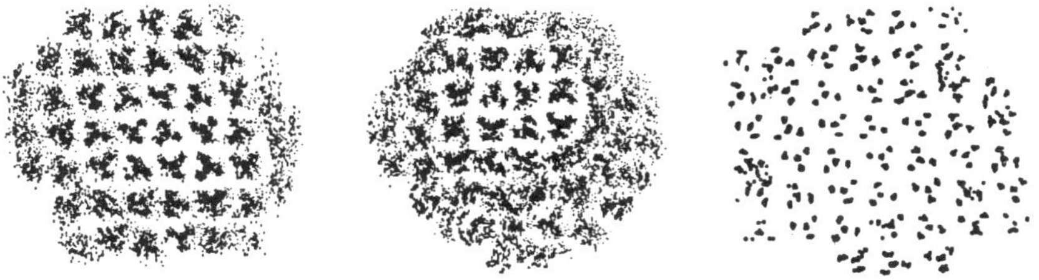

Figure 5.5 Images of a crystalline cluster of 188 molecules of *t*-butyl chloride undergoing a transition from tetragonal to monoclinic. Surface molecules are disordered. The outer phase also begins to disorder as the more compact inner phase grows

points, one dot for each bond directed toward the hemisphere, including dots from all molecules in the cluster, is a Pawley projection. Most commonly an 'equal area' plot is made in which each area of the plot corresponds to the area of the hemisphere from which the projections were made, preserving the azimuthal angles of the dots relative to the polar axis of the hemisphere. Typical results are shown in Figure 5.6 for a cluster of the octahedral molecule SeF_6 transforming from a body-centered cubic phase to monoclinic [23]. For the hexafluorides there are over a dozen stable or metastable phases, not all of which give different Pawley plots. Nevertheless, plots of those phases encountered most frequently differ markedly from each other and offer a rapid identification of the phase.

Most MD investigations of clusters in the past have been carried out on atomic clusters. The standard diagnostic functions or indices invoked to analyze or characterize the phases encountered include the pair correlation function $g(r)$, the temperature dependence of the Lindemann index δ, the coefficient of translational diffusion D_r, and the angular distribution function $P(\theta)$. For clusters of *polyatomic* molecules all of these diagnostics can play a role but the additional aspect of molecular orientation arises that has no counterpart in atomic clusters.

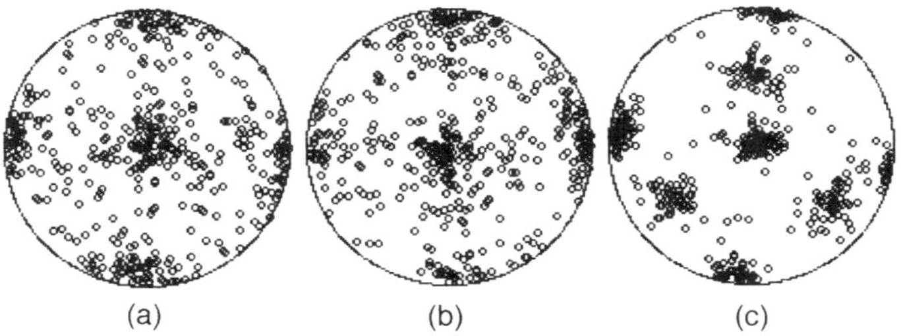

Figure 5.6 Pawley projections of molecules in an $(SeF_6)_{150}$ cluster. (a) b.c.c., 88 K; (b) at beginning of transition, 78 K; (c) monoclinic, 68 K. Note the substantial orientational disorder associated with the plastically crystalline b.c.c. structure

Orientational functions $g_\theta(r)$, D_θ, $Q(\theta)$ and indices δ_θ analogous to those listed above have been defined and found to have some utility. Each different phase has its own characteristic $g(r)$, $g_\theta(r)$, $P(\theta)$, and $Q(\theta)$ functions but they have not been as delicate, informative, or as convenient in analyzing phases as the Lindemann index and the coefficients of translational and rotational diffusion. Therefore, rather than discussing them further, we simply refer the reader to the original literature [45].

The Lindemann index was first suggested on the basis of empirical evidence. For simple atomic or ionic solids it was proposed that, when the root-mean-square displacements, $[\langle(r - \langle r\rangle)^2\rangle]^{1/2}$, of the distances between the atoms or ions in contact with each other exceed a certain fraction of their mean separation $\langle r\rangle$, the crystalline structure can no longer maintain its integrity and it melts [46]. Molecular-dynamics simulations tend to bear this out, and δ, the ratio of r.m.s. amplitudes to mean separations is of the order of magnitude of 0.1 at the transition.

In the case of small polyatomic molecules, if the separations of centers of mass are considered instead of separations between contacting atoms, the threshold ratio tends to be smaller, perhaps in the vicinity of 0.08. The advantage of the Lindemann index is that it is supposed to be more sensitive to transitions than the inflections in $U(T)$, and is intrinsically less noisy than the diffusion coefficients. On the other hand, it is not a primary indicator of transition, and its threshold value depends upon the particular system it is applied to. Therefore, it must be calibrated, somehow, either against $U(T)$, which would defeat the purpose of sensitivity, or against other indicators such as the coefficient of diffusion which *is* a primary indicator. Free self-diffusion is the essential characteristic of the liquid phase. But self-diffusion is a biased random walk and, as such, is intrinsically very noisy, requiring averaging over as many independent time origins and as many molecules as possible (recognizing the desirability of being able to distinguish between molecules at different sites in a cluster). Also, because the outer layers of molecules in clusters exhibit unusually high diffusion even when the clusters are crystalline according to most indicators, not even self-diffusion is free from arbitrariness as a criterion.

From the above discussion it is evident that experience and careful judgment are needed to interpret the phase behavior of clusters. Before going on to the next section it is worthwhile to illustrate the effect of size on transition temperature and on the diffuseness of transitions in small clusters. This is done in Figures 5.7 and 5.8 where the lack of sharpness and decrease in transition temperature as the size decreases are conspicuous [45, 47]. More will be said about these properties later.

The above diagnostic procedures are intended for gross identification of phases. They are all inadequate for detecting individual molecules or small nuclei in the process of forming a new phase in the presence of large numbers of molecules of the old phase. A diagnostic test found to be successful in

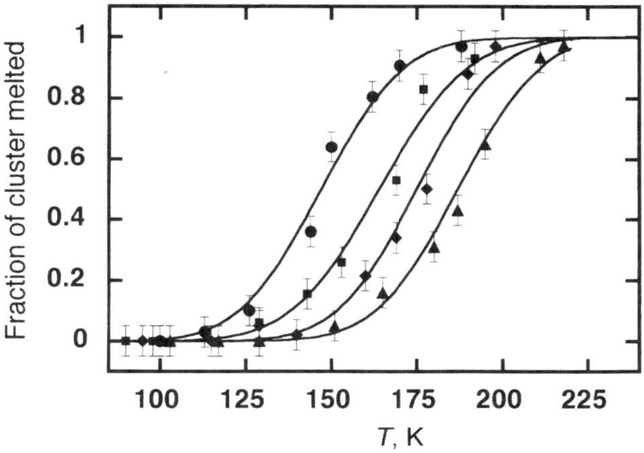

Figure 5.7 Fraction of molecules in the liquid state in individual $(TeF_6)_N$ clusters as they are heated. From left to right, $N = 100$, 150, 250, and 350

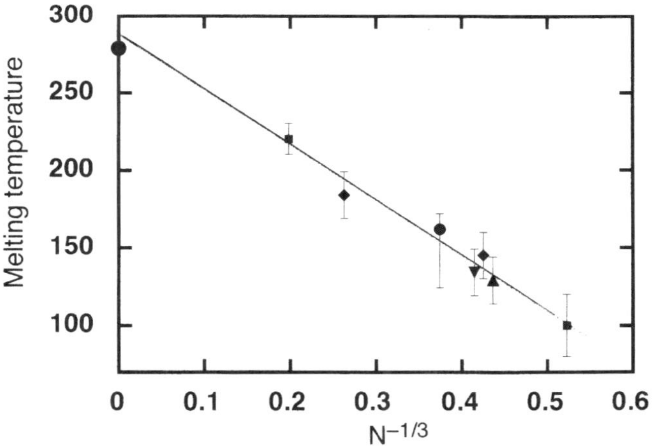

Figure 5.8 Midpoints of melting transitions of $(C_6H_6)_N$ clusters according to Monte Carlo simulations. Classical capillary theory predicts a linear plot for T_m against $N^{-1/3}$. The line is a least squares fit of the cluster points. On the ordinate is shown the experimental melting point of bulk benzene. Cluster points: $N = 7$, 12, 13, 14, 19, 55, 128. Note effect of the magic number stability at $N = 13$

accomplishing this task in atomic systems has also proven to be satisfactory in molecular systems. It is based on the computation of Voronoi polyhedra for each of the molecules present [48]. The faces of a Voronoi polyhedron for any given molecule (not on a surface) are segments of planes bisecting the lines connecting that atom with all of its neighbors. Planes for atoms too remote do not intersect the polyhedron and are irrelevant. Therefore, it is appropriate to introduce cutoff distances beyond which faces need not be considered.

A given polyhedron is characterized by the number of its faces and the number of edges possessed by each face. Typical liquids and various crystalline phases are each distinguished by their unique distributions of polyhedra. For example, all molecules in an ideal b.c.c. crystal have polyhedra consisting of 14 faces, 6 with 4 edges, and 8 with 6 edges. For molecules in a vibrating lattice, a few other distinct polyhedra occur with small probabilities. Polyhedra of molecules in a liquid rarely correspond to those in b.c.c. crystals, and if it is required that the b.c.c. polyhedra of at least two molecules must share a face to be considered as belonging to a b.c.c. nucleus, the chance of misidentification is small. The diagnostic criterion has been found to work well in detecting transient solid nuclei even smaller than critical nuclei in molecular liquids of the hexafluorides and might turn out to be useful in other types of systems [49]. It does not work effectively in the well-studied solid-state transition from b.c. to monoclinic in hexafluoride clusters, however, so that an alternative method must be devised, as outlined below.

The reason Voronoi polyhedra are ineffective in recognizing monoclinic nuclei forming from supercooled b.c.c. crystals is the reason the transformation is fast enough to take place on the MD time-scale in the first place, namely because molecules undergo very little translational reorganization. The principal change is for one-third of the quasispherical molecules to rotate by 60°. Therefore, the Voronoi polyhedra are little changed by the transformation. What is needed is a test to match individual molecular orientations with a reference orientation. One such test defines the index [45]

$$S_{ij}(\mathbf{r}, n) = (1/N)\sum\sum[(\mathbf{n}_{\alpha i}\cdot\mathbf{n}_{\beta j} + 1)/2]^n \tag{8}$$

where the sum is over pairs of equivalent bonds, N in number with unit vectors $\mathbf{n}_{\alpha i}$ and $\mathbf{n}_{\beta j}$ along their directions, between a molecule i to be tested and a reference molecule j. A perfect match gives unity for the result, while parameter n controls the latitude tolerated in the matching.

5. ADVANTAGES AND LIMITATIONS OF CLUSTERS AS MODEL SYSTEMS

5.1 Advantages

As reviewed above, interiors of molecular clusters quite faithfully model bulk matter and techniques applicable to clusters offer an alternative avenue in the search for ways to study condensed matter in transition. A troublesome aspect of conventional approaches to measuring nucleation rates has been the difficulty of adequately freeing systems from heterophase contaminants, particularly because the subdivided particles required by the technique must be separated mechanically by foreign, and not assuredly inactive, barriers [10]. Clusters, whether generated in supersonic flow or in computers, avoid the problem.

Besides making it possible to work with systems free from the influence of contamination, several other potential advantages are associated with clusters. As a consequence of the small volumes and rapid cooling rates that can be imposed, exceptionally deep supercoolings can be attained. In many cases this supercooling leads to extraordinarily high nucleation rates, rates which can easily be followed by the techniques applicable to clusters. This allows nucleation to be examined in heretofore unstudied regimes, and sometimes leads to the formation of phases never before seen in the bulk [13, 50]. In addition, both experimental and computational techniques are as applicable to certain solid-state transformations as to freezing. They have yielded the first homogeneous nucleation rates known to the writer for solid-state transitions in pure, one-component systems [26]. Especially important among the results yielded by the molecular-dynamics simulations are the direct views of the cooperative motions of the molecules. Critical nuclei, entities that until recently had been more the hypothetical inventions of theorists than the concrete objects of observation, can be watched as they form.

5.2 Limitations

Favorable as the above advantages of clusters may be, they do not come without costs. It is not possible by current techniques to control the temperatures of freely coasting clusters over wide ranges. Generally temperatures are in the vicinity of the so-called evaporative cooling temperature, as discussed in previous section. In addition, with present experimental arrangements, the time scale of experiments is limited to the range from a microsecond to perhaps one hundred microseconds. If it were expedient to do it, this range could be extended appreciably in either direction. A consequence of the thermal and temporal limitations is that only a fraction of possible transitions are accessible to the current experimental techniques for clusters. Even more limited are the time scales of MD simulations feasible with today's computer hardware, although the temperature can, of course, be controlled as desired.

How large nucleation rates must be if transitions are to take place in clusters during experiments or computations can be inferred directly from the characteristic cluster sizes and time scales of the method of investigation. Nucleation rate J is defined to be the number N^* of critical nuclei produced per unit volume per unit time, or

$$J = (1/V)\,\mathrm{d}N^*/\mathrm{d}t. \tag{9}$$

From the definition it is elementary to estimate what magnitude J must have if a critical nucleus is likely to form in the volume of a cluster during the time available in the experiment or simulation. In electron diffraction experiments with clusters of 100 Å and times of 10–100 μs, J must be of the order of $10^{28}-10^{30}$ m^{-3} s^{-1}. In MD simulations with cluster diameters typically about 30 Å and times of perhaps tens to hundreds of picoseconds at a given temperature, J must be about $10^{35}-10^{36}$ m^{-3} s^{-1}. When it is recalled that nucleation rates observed in conventional experiments on the freezing of small droplets are typically $10^{7}-10^{16}$ m^{-3} s^{-1}, or some 15 to 30 orders of magnitude lower than rates corresponding to the supersonic jet or the MD techniques, it is clear that it cannot be taken for granted that the transitions observed by the new methods can be considered to relate to the same process. Nevertheless, as will be seen, it has been possible to study some of the same transitions by molecular-dynamics simulations [21–23] that were observed in clusters by electron diffraction, despite the million-fold difference in rates. Unfortunately, only one transition, so far (the freezing of water), has been monitored by both the supersonic jet method [28] and by the conventional method [51–53] (at rates twenty orders of magnitude lower). Although the overlap of substances studied by the conventional method and by the cluster method has been regrettably small, the two methods gave results consistent with each other when interpreted in terms of nucleation theory. Therefore, there exist grounds for optimism that the cluster approach does provide information of relevance to bulk systems.

Most of the nucleation rates measured to date in clusters have been on materials so volatile that their vapors can be seeded into supersonic flows at room temperature. This situation is changing rapidly, however, as sample systems capable of operating at high temperatures are being designed.

Another concern about the cluster method is that not everyone agrees that matter confined to spaces as small as cluster volumes is genuinely representative of bulk matter [54]. These reservations are particularly strongly expressed in the case of supercooled water whose unusual properties are at the heart of an active controversy [55, 56]. Even in systems with more prosaic properties, the smallness of cluster volumes raises questions. The surface layer of molecules of a cluster constitutes a large fraction of the total volume. A simple estimate of the surface fraction F for a spherical cluster of N molecules is

$$F = 3(4\pi/3N)^{1/3}[1 - 0.5(4\pi/3N)^{1/3}]^2 \qquad (10)$$

For clusters of the sizes typically encountered in electron diffraction (N from about 10^4 to 10^3) and in MD simulations (N from about 500 to 50) these fractions are 21% to 41%, and 48% to 80%, respectively. Surface layers are not adequately taken into acount by the conventional capillary theory of nucleation. Quite apart from that problem, when the old phase wets the new phase being formed (a circumstance encountered in all of the systems to be described) nucleation in cooling clusters always occurs in the interior, and never in the surface layer. Therefore, in transitions occurring in MD simulations of clusters of a few hundred molecules, only a small fraction of the volume of the cluster can be considered to be effective in nucleation. In determining nucleation rates in clusters it is our convention to take the volume available for nucleation to be the fraction $(1 - F)$ of the total volume according to Eq. (10).

6. ASPECTS OF CLASSICAL NUCLEATION THEORY AND ITS PARAMETERS

6.1 Nucleation Theory for Condensed Phases

As initially formulated for freezing by Turnbull and Fisher [8], and Buckle [9], the steady-state rate of formation J of critical nuclei per unit volume per unit time was expressed as

$$J = A \exp(-\Delta G^*/kT) \tag{11}$$

where ΔG^* is the free-energy barrier to the formation from the liquid of the critical nucleus believed to initiate the transition. The prefactor in the Turnbull–Fisher theory is

$$A = N_v(kT/h) \exp(-\varepsilon/kT) \tag{12}$$

where N_v and ε represent, respectively, the number of molecules per unit volume, and the activation energy for the jumps of molecules across the boundary from the old phase to the new. A variant of the prefactor based on viscous flow theory to model the molecular jumps across the solid–liquid interface in the classical theory of homogeneous nucleation, uses [24]

$$A = 2(\sigma_{sl} kT)^{1/2}/(v_m{}^{5/3}\eta) \tag{13}$$

where σ_{sl} is the interfacial free energy per unit area of the boundary between solid and liquid, v_m is the molecular volume $1/N_v$, and $\eta(T)$ is the liquid viscosity. Equation (13) was applied in the earlier cluster reports. Several lines of evidence now suggest that this prefactor considerably exaggerates the inhibition of nucleation by the high viscosity at low temperatures. Better prefactors appear to be a heuristic 'free jump' prefactor [13] and one giving rather similar results but based on a more rigorous foundation, the prefactor of Grant and Gunton [57]. The latter is given by

$$A = \kappa\Omega/V \tag{14}$$

where, for a system of volume V, κ is the dynamic prefactor

$$\kappa = 2\lambda\sigma_{sl}T/LR^{*3} \tag{15}$$

with λ, L, and R^* representing the thermal conductivity, heat of fusion per unit volume of solid, and radius of the critical nucleus. Nuclei smaller than the critical size are unstable and tend to melt whereas larger ones grow spontaneously. In Eq. (14) Ω is the dimensionless statistical prefactor

$$\Omega = (2/27^{1/2})(V/\xi^3)(\sigma_{sl}\xi^2/kT)^{3/2}(R^*/\xi)^4 \tag{16}$$

with ξ representing a correlation length characterizing the thickness of the interface between the phases. In the classical theory R^* is given by

$$R^* = -2\sigma_{sl}/\Delta G_v \tag{17}$$

where $\Delta G_v < 0$ is the Gibbs free energy of the transition from the old to the new phase per unit volume at the supercooled temperature calculable from the relation

$$\Delta G_v(T) = -\int_T^{T_m} \Delta S(T)_V \, dT \tag{18}$$

In the above expression $\Delta S(T)_V$ is the entropy change per unit volume at temperature T estimated with the aid of an extrapolation of the difference between the heat capacities of the two phases.

A related formulation can be applied to solid-state transitions. For those so far encountered in clusters, the molecular jumps to the new phase correspond principally to reorientations instead of to translations, which is why the transitions are facile enough to be seen by our supersonic technique in the first place. For such cases the pre-exponential factor takes the form [26]:

$$A = 2(\sigma_{ij}kT)^{1/2}v_m^{-2/3}[\nu_0 \exp(-\varepsilon/kT)] \tag{19}$$

where the rotational jump frequency $[\nu_0 \exp(-\varepsilon/kT)]$ can be estimated from spectroscopic information or rotational diffusion coefficients derived from molecular-dynamics simulations [22, 23].

The free-energy barrier, ΔG^*, to the formation of a spherical nucleus is given by

$$\Delta G^* = 16\pi\sigma_{ij}^3/[3(\Delta G_v + w')^2] \tag{20}$$

as shown by Gibbs over a century ago [3]. The term w' may include the work of elastic deformation of a nucleus growing in a solid phase or the work of changing the surface area of an outer liquid phase i due to the volume change accompanying the formation of the nucleus of phase j. Some evidence suggests that the contribution to ΔG^* from the misfit between the old solid phase and the new may play a minor role in clusters [26], but whether this is true remains to be seen. It does appear, however, that the frequency in Eq. (19) that is satisfactory for nucleation is too low to account for the subsequent growth rate. The contribution w' for a nucleus growing in a spherical liquid of radius r_0 is $(2\sigma_i/r_0)(\rho_i - \rho_j)/\rho_i$, where $2\sigma_i/r_0$ is the Laplace pressure exerted by the outer phase on the inner phase and the ρ's are densities of the phases [58]. In experiments, it has been in the exponent $\Delta G^* / kT$ rather than in the factor A that σ_{ij} exerts the overriding effect on the nucleation rate. For example, a 20% change in σ_{sl} for CCl_4 changed the calculated J for freezing at 175 K by 10 orders of magnitude [24]. A similar sensitivity in MD simulations has not yet been encountered because at the extreme supercooling required to initiate nucleation, ΔG^* is lowered so greatly that the rate begins to approach the magnitude of the prefactor.

From the above relations it is apparent that the parameter σ_{ij} can be derived from the observed nucleation rate if the temperature of nucleation and the other physical properties in the equations are known (or can be plausibly estimated). For sake of example we shall adopt Eqs (14) and (20) and the classical expressions for ΔG^* and R^* but we emphasize the desirability of development of alternative, more rigorous treatments. Some aspects of flaws in the classical theory are discussed in the next subsection.

6.2 Remarks about the Free-Energy Barrier

In the studies of nucleation rates it has been customary to interpret results in terms of the classical (capillary) theory of nucleation sketched in the preceding paragraphs. The parameter in this theory about which least is known is the 'interfacial free energy' σ_{ij}. Derivable from the nucleation rate, σ_{ij} plays a role crudely analogous to that of the activation energy in the kinetics of chemical reactions in that it is a key source of the free-energy barrier to nucleation. Although its name is suggestive of a thermodynamic variable, it is, of course, a kinetic parameter. Its most important role, perhaps, is to facilitate projections of the nucleation rate at a given temperature to higher or to lower degrees of supercooling. Such estimates have proven to be helpful in forecasting whether or not transitions seen to occur on the microsecond time-scale of electron diffraction experiments might take place at greater supercoolings on the picosecond time-scale of molecular-dynamics simulations. To what extent σ_{ij} reflects the true thermodynamic variable has not been determined very precisely, partly because of the considerable difficulties associated with measurements of the thermodynamic variable.

Additional factors confuse the interpretation of σ_{ij}. Because it is possible for two phases to coexist at equilibrium at ambient pressure at only a single temperature, some theorists regard it as meaningless to consider σ_{ij} at any other temperature [59]. Worse, as already mentioned, the classical theory of nucleation has only qualitative validity and, therefore, the kinetic parameter σ_{ij} cannot be a true physical variable. The two most serious deficiencies of the capillary treatment for ΔG^* are the complete neglect of the thickness of the interface between the old phase and the critical nucleus, and the attribution of bulk-like properties to a critical nucleus containing only a handful of molecules. Serious though these deficiencies may be, we will interpret results in terms of conventional capillary theory for the remainder of the chapter because it does provide a means of comparing results of enormously disparate nucleation rates on a common and intuitive basis. Moreover, no generally applicable and accepted alternative exists.

It has been proposed that σ_{ij} is mildly temperature-dependent. In the case of the solid–liquid interface of metals, at least, it was suggested by Turnbull [60] that there is a tendency of the liquid to become more ordered at the interface to fit closely against the surface of the solid. This drive to fill space makes the interfacial entropy negative, tending to make σ_{sl} *increase* as the system warms. In the following it will be assumed that

$$\sigma_{\text{sl}}(T) = \sigma_{\text{sl}}(T_1)(T/T_1)^{\nu} \tag{21}$$

where ν is a number whose absolute magnitude is expected to be rather less than unity.

One way of assessing the plausibility of values of interfacial free energies derived from rates of nucleation is to see if they are more or less consistent with an empirical relation found by Turnbull [61], who noted that σ_{sl} as determined from freezing rates tends to be a fairly universal fraction, k_T, of the enthalpy of fusion per unit area, or

$$\sigma_{\text{sl}} \approx k_T \Delta H_{\text{fus}}/(V^2 N_{\text{A}})^{1/3} \tag{22}$$

where ΔH_{fus} and V represent, respectively, the molar enthalpy of fusion and molar volume, and N_{A} is Avogadro's number. For metals k_T is about 0.45 and for semimetals and nonmetals it is often approximately 0.32. We shall see that an analogous relation appears to apply to solid-state transitions.

7. ILLUSTRATIVE RESULTS

7.1 Experimental

7.1.1 Systems Amenable to Investigation

Rates of both freezing and solid-state transitions have been measured by scattering electrons from clusters in supersonic jets. Earlier investigations of dozens of substances had found systematic relations between thermodynamic properties of a material and the phase produced in a supersonic flow [19]. Specifically, it was discovered empirically that if a dimensionless index R_{c} based on the normal boiling point T_{b}, melting point T_{m}, and entropy of fusion ΔS_{m}, was defined to be

$$R_{\text{c}} = (T_{\text{b}} - T_{\text{m}})/T_{\text{b}} + 0.007(\Delta S_{\text{m}}/R)^2. \tag{23}$$

then, when the index for a substance exceeded (or fell short of) $R_{\text{c}} \approx 0.32$, the clusters obtained would be liquid (or crystalline). If, however, the index for a material were close to the threshold value of 0.32, clusters could be produced in either state, depending on expansion conditions. Therefore, it was possible to anticipate systems likley to be convenient candidates for studies of nucleation rates. After the criterion for predicting cluster phase via R_c had been formulated empirically and tested, the kinetic rationale underlying its success was recognized and found to pertain to clusters with diameters in the vicinity of 100 Å monitored for a period of some dozens of microseconds [62]. By working with clusters appreciably larger or smaller, and by changing the spatial dimensions and the duration of the experiment it is possible to alter the threshold value somewhat from 0.32. No comparable criterion has yet been developed for solid-state transitions but systems that undergo nonreconstructive transformations in the bulk at temperatures appreciably above the evaporative cooling temperature are possible candidates.

7.1.2 Reasons for Studying Freezing

Nucleation rates are not particularly interesting pieces of information in themselves. Nevertheless, several questions involving them do arise and these deserve answers. What is notably different about the cluster technique in nucleation research is the considerably smaller size of the droplets subjected to examination. This smaller size together with the extremely short times of observation require nucleation rates to be truly enormous if freezing is to be seen. Therefore, liquid clusters attain a much greater supercooling before they freeze than droplets did in previous studies. Are any new phenomena observed in this new regime of supercooling? Do the astronomically greater rates encountered imply a fundamentally different mechanism of freezing? Are the liquid–solid interfacial free energies derived for the rates in accord with the Turnbull relation (Eq. 22)? Are the crystalline phases encountered the same as those found in studies governed by thermodynamics? Preliminary answers have been found for all of the questions.

7.1.3 Freezing Results

Cluster systems whose nucleation rates have been measured include carbon tetrachloride [24], trichlorobromomethane [63], 1,1,1-trichloroethane [25], 2,2,3-trimethyl butane [63], cycloheptane [63], dimethyl acetylene [29], water [28], ammonia [27], and phosphorus trifluoride [63]. It was found that the crystalline phases formed for carbon tetrachloride, trichlorobromomethane, dimethyl acetylene, and water were not the phases obtained when the liquids freeze slowly at their freezing points. Not enough is known about trimethyl butane to draw a conclusion. Bulk carbon tetrachloride exhibits a complex behavior. When the liquid is cooled it solidifies to a metastable cubic form (Ia) which only slowly transforms to the stable rhombohedral form (Ib) unless the system is cooled below 234 K at which temperature the transition to Ib becomes rapid. Below 225 K the stable structure is monoclinic (II). Our liquid clusters freeze rapidly at 175 K directly to the rhombohedral form. Dimethyl acetylene freezes into an unidentified form unknown in the bulk. Water clusters freeze into cubic ice (Ic), not the stable hexagonal ice (Ih). Because water is the most intensively studied of all materials, it is possible to provide a more detailed account of its behavior than for the other examples. Moreover, its example affords a simple explanation for the formation of a metastable rather than a stable phase in the nonequilibrium process.

Despite the enormous attention paid to water, the anomalous properties of the supercooled liquid remain inadequately explained [55, 56], and unresolved controversies over the interpretation of these properties remain. What is clear from our experiments is that 75 Å clusters (about 6000 molecules) cooled to 200 K manage to avoid the critical instability hypothesized to occur in bulk water in the vicinity of 228 K but heretofore untested because no technique existed to cool water below about 235 K, without freezing it, while it was being monitored. Whether this immunity of clusters is due to their small size or because the hypothetical instability is not real cannot yet be concluded definitively. Figures 5.2 and 5.3, which show the time evolution of electron diffraction patterns and of the temperature of water clusters as they cool, verify that clusters remain liquid, uneventfully, until they begin to freeze at about 200 K [28].

As mentioned above, water is of special significance because it is the only system so far to have been studied by both the conventional [51–53] and the cluster techniques [28]. Figure 5.9 compares our nucleation rates for clusters

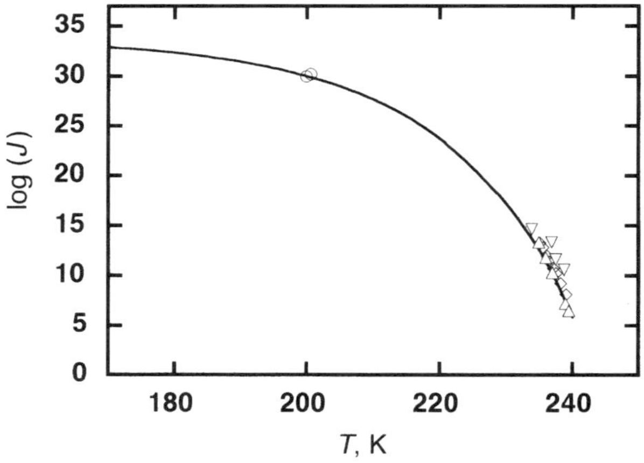

Figure 5.9 Logarithms of experimental nucleation rates for the freezing of water: clusters [28], circles; water-in-oil emulsions [51–53] triangles and diamonds. Solid curve, calculated temperature dependence according to Eqs (11), (14), (20), and (21) with $\nu = 0.07$

generated from runs with water vapor initially at 95 °C and 120 °C (seeded into 3.5 atm of neon carrier gas) with those from droplets dispersed in an emulsion [56, 57]. Despite the 15-to-25 order-of-magnitude difference in nucleation rates, the cluster and droplet results are naturally connected by nucleation theory, as suggested by the line constructed to pass through the cluster results and projected to lower supercoolings using a value of ν in Eq. (21) of 0.07. This conclusion is more easily recognized in Figure 5.10 which plots the interfacial free energies derived from the nucleation rates. For uniformity and simplicity, all of the kinetic parameters in the figure were recalculated directly from the reported nucleation rates and put on the same basis, using the Grant–Gunton prefactor A of Eq. (14). In the earlier work on droplets, the authors had adopted various and appreciably different conventions, especially in their simplified calculations of the free energy of freezing. What is of particular interest is the fact that the kinetic interfacial free-energy parameters for the clusters and for the droplets are in good agreement with each other but in considerable disagreement with the thermodynamic value derived from the equilibrium contact angle method at the equilibrium freezing point [64, 65].

The interpretation of these facts is quite straightforward. The ice in contact with the liquid is the normal, thermodynamically stable, hexagonal ice. The ice to which the liquid clusters froze was readily identified as cubic ice. Although the solid to which the liquid droplets crystallized in the experiments of Wood and Walton [51], of Butorin and Skripov [52], and of Taborek [53], was not determined, the present results leave no doubt that it was also cubic ice. Despite the claim of Wood and Walton [51] that their temperature-dependent interfacial

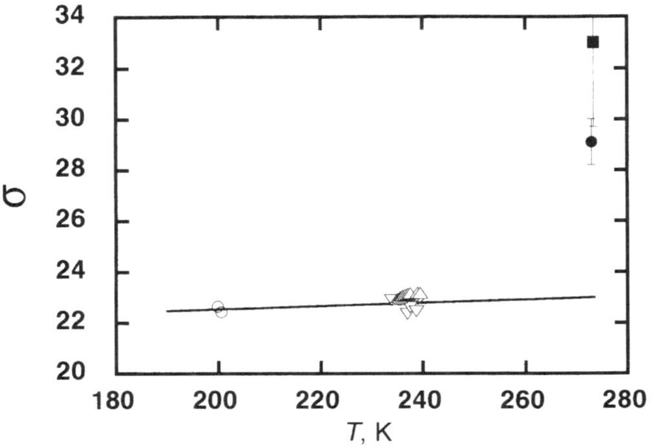

Figure 5.10 Experimental ice–water interfacial free energies from nucleation rates in clusters (circles) and emulsion experiments (triangles and solid diamonds), and from contact angles [64, 65] at 273 K (markers with error bars)

free energy extrapolated to that of hexagonal ice in water at the freezing point, the temperature dependence of their extrapolation was much too steep, and the real dependence was buried in the experimental noise [28, 52]. As confirmed by the low interfacial free energy derived from the nucleation rates, the reason that cubic ice forms at low temperatures instead of the stable hexagonal ice, is kinetic. The low interfacial free energy for cubic ice is associated with a much lower barrier to the formation of critical nuclei. The low interfacial free energy had, in fact, been anticipated on the basis of structural aspects of hydrogen bonding at the interface [66].

7.1.4 Discussion of Freezing Experiments

In the foregoing was answered the question of what governs the phases formed at great supercooling. It was seen that the kinetics of the process governs the product. The barrier to nucleation, arising from the interfacial free-energy cost of forming the liquid–solid boundary, determines what happens first. Sometimes, as in the case of ammonia, but not always, the result is the same as that expected from thermodynamics. Are the liquid–solid interfacial free energies derived from the rates in supersonic jets consistent with those derived from rates twenty orders of magnitude lower in macrosopic systems? For water this turned out to be true. What can be said about the other systems for which no experimental comparisons are available is that the derived interfacial free energies were all more or less in accord with Turnbull's empirical relation [61], and, hence, unexceptional in spite of the enormous nucleation rates and smallness of particle size. This is not particularly surprising, because the critical nuclei, whose interfaces are at issue, are still small in comparison with the clusters. Taking the above information into account, it does not appear that the astronomically higher rates in clusters are associated with a fundamentally different mechanism of freezing. The only phenomenon encountered in the new regime of supercooling that might be considered as new is the marked tendency for the formation of metastable rather than stable phases. It is not clear that even this behavior is particularly new. In the earlier work on nucleation rates it was not known what phase was first nucleated. In the few cases where the structure of the product was studied, the long times between the phase change and diffraction investigation would have been sufficient for the system to anneal to a more stable phase. If it were desired to retain and work with a metastable phase, this could probably be done by quenching.

7.1.5 Solid-State Transformations

When phase changes take place in solids, the process, in general, has been found to be brought about by heterogeneous nucleation. Insofar as the present author is aware, the first homogeneous nucleation rates measured for solid-state transitions in one-component systems were those for clusters in supersonic jets [26]. Conditions in the supersonic experiments that favor homogeneous nucleation are the freedom from heterophase impurities and grain boundaries (our clusters, individually, are each single crystals) and the extreme supercooling the clusters experience. The fact that phase changes in (rigid) solids can be competitive in speed with the freezing of liquids at the temperatures in supersonic jets (and, as we shall see, even faster than freezing at the higher supercoolings in MD simulations) is less anomalous than simple intuition might suggest. The only solid-state transitions observed so far in clusters have been nonreconstructive. That is, the principal reorganization involved has been a change from a disordered orientation to an ordered orientation of quasispherical or of quasi-cylindrical molecules. Little translational reorganization has been required.

The two crystalline cluster systems whose nucleation rates have been measured [26] are SeF_6 (b.c.c. to monoclinic) and *t*-butyl chloride (tetragonal to monoclinic). An incomplete study [67] indicates that *t*-butyl thiol undergoes a change akin to that of the closely similar *t*-butyl chloride. Because computer simulations provide even more information about the transformations of *t*-butyl chloride and the hexafluorides, we defer further discussion about solid-state transitions to the next section.

7.2 Computational

7.2.1 Some General Aspects of Transitions in Clusters

The first question that arises is how to make a plausible choice among all the potential candidates for the study of nucleation. Only a fraction of known phase transitions are accessible to simulation with presently available capabilities of computers. Our approach was to take clues from the successful supersonic experiments. It was possible to apply nucleation theory to project whether the nucleation rate would increase sufficiently with deep supercooling to make simulation feasible. In spite of the quantitative deficiencies of the classical theory, the theory did make projections of the temperature dependence of the rates accurately enough to indicate which of the transitions observed experimentally would increase the 6 or 8 orders of magnitude required to allow them to take place on the time scale of computations.

Next are discussed some properties of clusters of particular concern in MD simulations. Figure 5.4 shows how the configurational energies vary as SeF_6 clusters of various sizes are heated and cooled [23]. Except for sign, the potential energies correspond to the energies of sublimation, for the molecules are modeled as rigid entities. Several striking features deserve comment. First, clusters of different sizes have conspicuously different stabilities. Next, the transition temperatures indicated by inflections in the curves are not sharp and they depend on the cluster size. Last, the cooling curves do not follow the heating curves in the vicinities of the transitions. It turns out that heating curves are quite reproducible but the cooling curves are not. Phase changes on cooling require a nucleation event, and nucleation is a stochastic process. Therefore, the chance event by which structural fluctuations in the warmer phase lead to critical nuclei of the colder phase are by their nature unpredictable and may occur sooner or later at lower or higher supercoolings. Nucleation in the present systems (whose liquids wet the solids) *always* occurs in the interior of a cluster, *never* on the surface. On the other hand, when heated, crystals go from lower-symmetry structures with tighter, more rigid packing, to more disordered, more loosely packed structures with a higher (average) symmetry. When clusters are heated, the phase change always begins in the surface layer of molecules and progresses inwards as the cluster is warmed. It appears that the looser organization of the higher-temperature phase already exists in regions of the disordered surface of clusters. Therefore, no specific nucleation event is needed to obtain the warmer phase during heating runs, and no superheating is required. This is not true in MD runs imposing periodic boundary conditions where the system, being denied a surface, may need to be superheated considerably.

If small clusters differed intrinsically throughout their volume from large clusters, there would be no logical way to relate nucleation rates in them to the classical theory because there would be no way to decide the degree of supercooling. Fortunately, it is possible to account quite satisfactorily for the substantial differences in the caloric curves of Figure 5.4 by apportioning the configurational energy to a core energy and a Gibbs excess surface energy. When this is done in a straightforward way without any reference to the bulk [26] it is found that the core energy agrees quantitatively with the bulk energy calculated with periodic boundary conditions, and that the surface energy is quite normal. For this and other reasons to be related below, it can be assumed with some confidence that cores of clusters resemble bulk matter closely enough to apply nucleation theory with supercooling reckoned from the bulk transition temperature. The lack of sharpness of transitions is associated with the difference between the core and the surface layer of molecules, the surface comprising fully two-thirds of the volume of a 150-molecule cluster.

7.2.2 Freezing of Liquid Clusters

As mentioned in the experimental section, it has proven to be simpler to carry out molecular-dynamics simulations of certain solid-state phase transitions than of the freezing of supercooled liquids. The reason appears to involve both the greater amount of reorganization of molecular positions in freezing and the greater interfacial free energy (and hence, the greater barrier to nucleation). In runs an order of magnitude longer than those needed for the solid-state transitions it was possible to freeze molten 150-molecule clusters of the chalcogen hexafluorides SF_6, SeF_6, and TeF_6 [49]. The liquid starting points had been constructed by heating crystalline samples. Because each cooling run took months of CPU time, insufficient data were acquired to determine nucleation rates and the interfacial free energy σ_{sl} derivable from the rates. What was learned was that analyses of Voronoi polyhedra, as discussed in an earlier section, were effective in verifying the absence of crystalline seeds in the molten samples and in recognizing the formation of crystalline regions on a molecule-by-molecule basis, during cooling. When cooled at the rate of 2×10^{10} K/s the chalcogen hexafluorides all froze to single crystals, confirming that freezing was mononuclear. That is, freezing in each cluster was initiated by the formation of a single critical nucleus, a nucleus which then grew until the entire cluster was frozen.

7.2.3 Solid-State Transformations

In experiments [26] it had been shown that clusters of *t*-butyl chloride emerged from the supersonic nozzle as single crystals of the well-known tetragonal structure. This plastically crystalline phase III is stable in the bulk over the range from 218 to 183 K. At 156 K the supercooled clusters transformed to the thermodynamically stable phase IV whose structure was unknown. Although neutron-diffraction Debye–Scherrer powder patterns of quite high resolution existed, the patterns had not been satisfactorily indexed. After an application of nucleation theory indicated that the nucleation rate of greatly supercooled clusters would become high enough to be accessible to molecular-dynamics computations, simulations soon confirmed the prediction. Moreover, the form to which a 188-molecule tetragonal cluster spontaneously transformed when cooled was found to be monoclinic ($P2_1/m$). When the diffraction pattern calculated from the structure was found to agree quantitatively with the observed pattern, the result showed that MD runs are able to provide useful new structural as well as kinetic information [43].

Stages of the transition are shown in Figure 5.5. A series of cooling runs [22] established the nucleation rate for the transition. Analyses of the temperature-dependent coefficient of rotational diffusion, mean librational frequencies, and relaxation times of core molecules in the tetragonal phase agreed with a selected subset of the rather scattered experimental results from NMR, cold neutron diffraction, and dielectric techniques. This outcome, then, supported our assumption that cores of clusters behave sufficiently like the bulk to make it reasonable to apply nucleation theory to clusters. Moreoever, the results provided a temperature-dependent molecular jump frequency from the tetragonal to the monoclinic phase needed in the derivation of the interfacial free energy of the tetragonal–monoclinic boundary. It was found that the parameter σ_{sl} for the interface was about 3.5 mJ/m^2, or approximately 30% of the heat of transition for the phase change (per unit area of a molecular layer) just as Turnbull had observed in a series of freezing experiments of molecular liquids. Figure 5.11 compares the

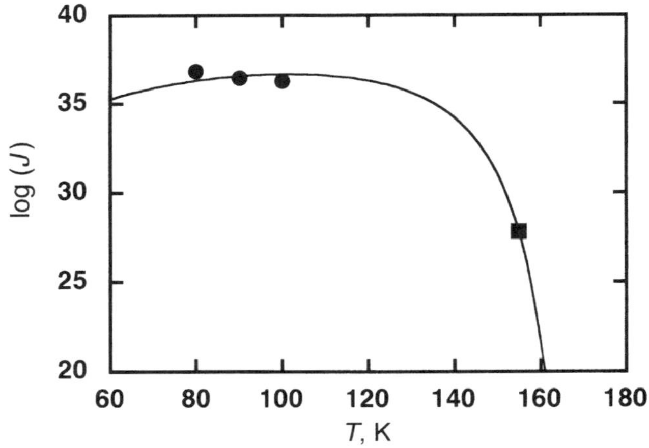

Figure 5.11 Logarithms of nucleation rates for the transition of *t*-butyl chloride from tetragonal to monoclinic. Circles, cluster simulations; square, supersonic cluster experiment; line, nucleation theory with constant interfacial free energy. Uncertainties, approximately the size of the markers

MD and experimental nucleation rates, connecting them via nucleation theory using a constant interfacial free energy [22]. At progressively greater supercoolings the nucleation rate first climbs steeply, then levels off as the free-energy barrier to nucleation becomes small and the prefactor decreases.

Similar results were found for transitions in a series of hexafluorides. Cores of clusters closely resembled the bulk materials, and interfacial free energies were roughly in accord with a generalization of the Turnbull relation, Eq. (22). What was different from the case of *t*-butyl chloride was the divergence of the kinetic outcome of the phase changes from the themodynamically favored result. All of the hexafluoride liquids lighter than the actinides freeze to plastically crystalline b.c.c. solids. As the lightest hexafluorides are cooled (SF_6 and, apparently, SeF_6, whose observed thermodynamically stable cold phase is inconclusively resolved [26] but is probably monoclinic) both thermodynamics and kinetics favor a transition to the monoclinic phase ($C2/m$, $Z = 3$). For the intermediate hexafluorides TeF_6, MoF_6, and WF_6, however, the stable cold phase is reported to be orthorhombic ($Pnma$, $Z = 4$) as it also is for the actinides. Clusters of the intermediate materials formed under conditions of cold flow in supersonic jets do indeed have this structure [50]. When supersonic flow conditions are warm, however (i.e. at higher subject/carrier ratios and lower pressures) the resultant intermediate clusters quickly transform from b.c.c. to the same monoclinic structure characteristic of the lighter hexafluorides. This structure is also formed in MD simulations of cooled b.c.c. clusters [45]. That it is metastable, not stable, is confirmed in MD runs comparing orthorhombic, monoclinic, and a number of other efficiently packed structures. Although the structures all have comparably low configurational energies with the orthorhombic being marginally lower, and all transform to b.c.c. when warmed, the decisive indication is the temperature at which the b.c.c. forms. The fact that the transition to b.c.c. occurs at a distinctly higher temperature for orthorhombic clusters than for the others [45] confirms that the orthorhombic form is thermodynamically more stable than the others. Whether orthorhombic is the stable phase at the lowest temperatures is uncertain because, under conditions of coldest flow in experiments, a new phase of MoF_6 and WF_6 has been produced in supersonic flow mixed with the orthorhombic [68]. It has not yet been identified. Conceivably it is the most stable phase at very low temperatures but is not seen in ordinary heat capacity or diffraction experiments on bulk materials because of the sluggishness of transition.

7.2.4 Validity of Simulations

Because molecular-dynamics simulations are extremely time-consuming, and more so the larger the system, it is fortunate that very small clusters of polyatomic molecules are able to model condensed matter with fair fidelity. Quite apart from considerations of the size of the system, the question arises as to how faithful the potential functions must be if meaningful results are to be obtained. Potential functions used in most of the simulations described in the foregoing were simple pairwise-additive, atom–atom functions based on 6-12 or 6-12-1 interactions. Three-body effects were neglected as were effects of polarizability when partial charges were included. It is well known that 6-12 functions, themselves, give a mediocre account of interatomic forces, particularly at short distances. The reason such simple, marginally accurate interactions were incorporated in the simulations was, of course, to reduce computational time to an acceptable level. Even with such simplifications, a single cooling run (with 10^5 time steps per 10 degree step) to achieve the freezing of a 150-molecule hexafluoride cluster devoid of partial charges takes several months of CPU time on an IBM RISC/6000-320 workstation. The practical question is not one of how to implement state-of-the-art potential functions but of how much complexity must be introduced to achieve a system that behaves in a realistic way. What is most important in studies of the present sort is whether the system reproduces the phase behavior characteristic of real systems. The empirical answer is that, for the hexafluorides and for *t*-butyl chloride, at least, the 6-12-1 functions parameterized to fit heats of sublimation yielded surprisingly good results. In the case of the quasispherical hexafluorides, even the 6-12 functions did a rather good job although the b.c.c.–monoclinic transition temperature was somewhat low when the partial charges were neglected. Even when the central atom was dispensed with, the molecules exhibited the observed phase behavior although the resultant intermolecular interaction gave a poorer account of the orthorhombic crystal structure [30]. For certain molecules such as N_2, the halogens, and acetylene, the 6-12 spherical atom approximation is unsatisfactory, and aspherical atoms are required to reproduce the structures of the various phases [69, 70]. For a discussion of such potential functions see the chapter in this volume by S. L. Price [69].

The great virtue of MD simulations, of course, is that they provide a view of processes at the molecular level that no experiments have been able to rival. Such poorly understood properties as the thicknesses of interfaces between phases can be analyzed. This information, which is latent within the computer outputs of existing runs, has yet to be harvested adequately. Inasmuch as these poorly understood properties are inadequately treated in the classical nucleation theory and may be of considerable importance, it is time that they should be illuminated. A significant step in this direction has recently been taken by Gránásy and coworkers [71].

8. CONCLUDING REMARKS

It is gratifying to find that the behavior of large molecular clusters in supersonic jets appears to correlate closely with the behavior of bulk matter in much more difficult experiments. Supersonic experiments, then, provide a new way to deduce information of importance in interpreting phenomena that have been troublesome to get at in the past. It is equally felicitous to find that phase changes simulated by molecular dynamics closely parallel those observed in supersonic experiments. This result lends credibility to conclusions about the molecular behavior derived from the simulations. Lest it be a source of concern that inferences about the classical MD trajectories of molecules violate the uncertainty principle, it turns out that the de Broglie wave lengths of the atoms in simulations are typically only a fraction of an atomic diameter and are comparable to amplitudes of oscillations of molecules in the condensed phase.

The idea of investigating the time evolution of the structure of molecular systems was a foreign concept in the field of structural chemistry until recently. It is becoming increasingly clear that kinetic aspects of structure are rewarding to study and lie at the very heart of chemistry.

9. ACKNOWLEDGEMENTS

I gratefully acknowledge helpful discussions with Professors D. W. Oxtoby, D. Turnbull, and E. J. Valente, and the chapter would have been much poorer without the skillful assistance of many people, including Drs. J. Huang and S. Xu, in particular. Molecular-dynamics calculations were performed using the CCP5 Program Library, Daresbury Laboratory, Daresbury, Warrington WA4 4AD (UK). I am indebted to Dr. W. Smith of the Daresbury Laboratory for the program MDMPOL used in the calculations.

10. REFERENCES

1. O. D. Sherby and J. Wadsworth, *Sci. Amer.,* **252**, 112 (1985).
2. J. P. Joule, *Philos. Mag.*, **27**, 205 (1845).
3. J. W. Gibbs, *Trans. Conn. Acad.*, **3**, 108 (1876); *ibid.* 343 (1878).
4. L. Farkas, *Z. Phys. Chem.*, **125**, 236 (1927).
5. M. Volmer and A. Weber, *Z. Phys. Chem.*, **119**, 277 (1926); M. Volmer, *Kinetik des Phasenbildung*, Steinhoff, Leipzig (1939).
6. R. Becker and W. Döring, *Ann. Physik*, **24**, 719 (1935).
7. A. Einstein, *Ann. Physik*, **33**, 1275 (1910).
8. D. Turnbull and J. C. Fisher, *J. Chem. Phys.*, **17**, 71 (1949).
9. E. R. Buckle, *Proc. R. Soc. (London)*, **A261**, 189 (1961).
10. D. Turnbull, *J. Chem. Phys.*, **18**, 198 (1950).
11. K. C. Young, *Microphysical Processes in Clouds*, Oxford University Press, New York (1993).
12. J. M. McBride, *Science*, **256**, 814 (1993).
13. L. S. Bartell, *J. Phys. Chem.*, **99**, 3924 (1995).
14. D. Turnbull, *J. Metals,* **188**, 1144 (1950).
15. D. Turnbull and R. E. Cech, *J. Appl. Phys.*, **21**, 804 (1950).
16. D. Turnbull, *J. Appl. Phys.*, **21**, 1022 (1950).
17. L. S. Bartell, L. R. Sharkey and X. Shi, *J Amer. Chem. Soc.*, **110**, 7006 (1988).
18. See, for example, L. S. Bartell, *Chem. Rev.*, **86**, 492 (1986).
19. L. S. Bartell, L. Harsanyi and E. J. Valente, *J. Phys. Chem.*, **93**, 6201 (1989).
20. See, for example, M. P. Allen and D. J. Tildseley, *Computer Simulations of Liquids*, Oxford Science Publications, Clarendon Press, Oxford (1987).
21. L. S. Bartell and S. Xu, *J. Phys. Chem.*, **98**, 6688 (1994).
22. L. S. Bartell and J. Chen, *J. Phys. Chem.*, **99**, 11147 (1995).
23. L. S. Bartell and S. Xu, *J. Phys. Chem.*, **99**, 10446 (1995).
24. L. S. Bartell and T. S. Debble, *J. Phys. Chem.*, **95**, 1159 (1991).
25. T. S. Dibble and L. S. Bartell, *J. Phys. Chem.*, **96**, 2317 (1992).
26. T. S. Dibble and L. S. Bartell, *J. Phys. Chem.*, **96**, 8603 (1992).
27. J. Huang and L. S. Bartell, *J. Phys. Chem.*, **98**, 4543 (1994).
28. J. Huang and L. S. Bartell, *J. Phys. Chem.*, **99**, 3924 (1994).
29. J. Huang, W. Lu and L. S. Bartell, *J. Phys. Chem.*, **99**, 11147 (1995).
30. K. E. Kinney, Ph.D. Thesis, University of Michigan (1995).
31. P. Audit, *J. Phys. (Paris)*, **30**, 193 (1969); B. Raoult and J. Farges, *Rev. Sci. Instrum.*, **44**, 430 (1973).
32. P. Wegener in *Nonequilibrium Flows*, edited by P. Wegener, Marcel Dekker, New York (1969), Ch. 4, p. 119.
33. R. Beck, M. F. Hineman and J. W. Nibler, *J. Chem. Phys.*, **92**, 7068 (1990).
34. J. D. Ewbank, D. W. Paul, L. Schafer and L. Bakhtiar, *Appl. Spectrosc.*, **43**, 415 (1989).
35. J. Farges, M. F. de Feraudy, B. Raoult and G. Torchet, *Surf. Sci.*, **106**, 95 (1981).
36. J. Gspann in *Physics of Electronic and Atomic Collisions*, edited by S. Datz, North-Holland Press, New York (1982).
37. C. E. Klots, *Phys. Rev.*, **A39**, 339 (1989).
38. L. S. Bartell, *J. Phys. Chem.*, **94**, 5102 (1990).
39. C. E. Klots, *Z. Phys. D.*, **20**, 105 (1991).
40. J. W. Hovick and L. S. Bartell, *J. Molec. Struct.*, **346**, 321 (1995).
41. See B. Raoult, J. Farges, M. F. de Feraudy and G. Torchet, *Z. Phys. D.*, **12**, 85 (1989); *Philos. Mag.*, **B60**, 881 (1989).

42. P. Buffat and J. -P. Borel, *Phys. Rev.*, **A13**, 2287 (1976).
43. J. Chen and L. S. Bartell, *J. Phys. Chem.*, **97**, 10645 (1993).
44. A. H. Fuchs and G. S. Pawley, *J. Phys. France*, **49**, 41 (1988).
45. S. Xu and L. S. Bartell, *J. Phys. Chem.*, **97**, 13544, 13550 (1993).
46. F. A. Lindemann, *Phys. Z.*, **11**, 609 (1910); *Engineering*, **94**, 515 (1912).
47. F. J. Dulles, Ph.D. Thesis, University of Michigan (1993).
48. M. Tanemura, Y. Hiwatari, T. Ogawa, N. Ogita and A. Ueda. *Prog. of Theor. Phys.*, **58**, 1079 (1977).
49. K. E. Kinney, S. Xu and L. S. Bartell, *J. Phys. Chem.*, **100**, 6935 (1996).
50. L. S. Bartell, J. W. Hovick, T. S. Dibble and P. J. Lennon, *J. Phys. Chem.*, **97**, 230 (1993).
51. G. R. Wood and A. G. Walton, *J. Appl. Phys.*, **41**, 3027 (1970).
52. G. T. Butorin and V. P. Skripov, *Sov. Phys. - Crystallogr. (Engl. Transl.)*, **17**, 322 (1972).
53. P. Taborek, *Phys. Rev.*, **B32**, 5902 (1985).
54. G. W. Robinson and S. B. Zhu in *Reaction Dynamics in Clusters and Condensed Phases*, edited by J. Jortner, Kluwer Academic, Amsterdam (1994), P. 423.
55. R. J. Speedy and C. A. Angell, *J. Chem. Phys.*, **65**, 851 (1976); C. A. Angell, *Ann Rev. Phys. Chem.*, **34**, 593 (1983).
56. Y. Xie, K. F. Ludwig, Jr., G. Morales, D. E. Hare and C. M. Sorensen, *Phys Rev. Lett.*, **71**, 2050 (1993).
57. M. Grant and J. D. Gunton, *Phys. Rev.*, **B32**, 7299 (1985).
58. N. Fukuta in Lecture Notes in Physics, *Atmospheric Aerosols and Nucleation*, edited by P. E. Wagner and G. Vali, Springer-Verlag, Berlin (1989), p. 504.
59. P. Harrowell and D. W. Oxtoby, *J. Chem. Phys.*, **80**, 1639 (1984).
60. D. Turnbull in *Physics of Non-Crystalline Solids*, edited by J. A. Prins, North-Holland, Amsterdam (1964), p. 4.
61. D. Turnbull, *J. Appl. Phys.*, **21**, 1022 (1950).
62. L. S. Bartell, *J. Phys. Chem.*, **96**, 108 (1992).
63. J. Huang, W. Lu and L. S. Bartell, unpublished research.
64. P. V. Hobbs and W. M. Ketcham, in *Physics of Ice*, edited by N. Riehl, B. Bullemer and H. Engelhardt, Plenum Press, New York (1969), p. 95.
65. S. C. Hardy, *Philos. Mag.*, **33**, 471 (1977).
66. T. Takahashi and T. Kobayashi, *J. Cryst. Growth*, **64**, 593 (1983).
67. J. W. Hovick, Ph.D. Thesis, University of Michigan (1994).
68. J. W. Hovick and L. S. Bartell, unpublished research.
69. S. L. Price, Chapter 2, this volume.
70. P. Santikary, unpublished research.
71. L. Gránásy, *Mater. Sci. Eng.*, **A178**, 121 (1994); L Gránásy, T. Pusztai and E. Hartmann, *J. Cryst. Growth*, in press.

Chapter 6

Ab Initio Prediction of Possible Molecular Crystal Structures

ROBERT J. GDANITZ
Fachbereich Physik der Gesamthochschule Kassel

In a *Nature* article of 1988, Maddox [1] wrote: 'One of the continuing scandals in the physical sciences is that it remains in general impossible to predict the structure of even the simplest crystalline solids from a knowledge of their chemical composition'. In 1990, this opinion was confirmed by Hawthorne [2]: 'Thus a rigorous general solution to the question of crystal structure prediction may not be forthcoming in the near future'. Concerning molecular crystals, Fagan and Ward [3] in 1992 expressed the opinion that 'Unfortunately the molecular interactions in a lattice become too complicated for any human or computer to handle'. At about the same time, the first papers [4, 5] successfully dealing with this problem appeared in the literature, disproving the statement of Fagan and Ward. Meanwhile, quite a few articles on predicting molecular crystal structures have been published, showing that this problem has now become a rather routine task. The present contribution deals with the various methods that have so far been published.

Theoretical Aspects and Computer Modeling Edited by A. Gavezzotti

1. PRELIMINARIES

1.1 What Is Meant by 'Predicting Molecular Crystal Structures'?

Chemists frequently 'create' molecules on a purely theoretical basis by specifying what kind of atoms they contain and which atom pairs are connected by chemical bonds. Being able to obtain the equilibrium structure of a molecule (the standard methodology of quantum chemistry is actually capable of this, provided the molecule is not too large) we may ask questions as to how an infinite number of these molecules will arrange themselves to build a crystal? To answer this question (again on a purely theoretical basis) the so-called *packing problem* has to be solved.

1.2 Obstacles of the Packing Problem

Recently, Gavezzotti [6] wrote a review with the title 'Are crystal structures predictable?' and giving 'no' as the answer. Indeed, there are obstacles (some of them being unavoidable in a practical sense, e.g. those due to polymorphism) that prevent crystal-structure predictions from being straightforward computations that take a fixed amount of time, depending on the accuracy one wants to obtain.

1.2.1 Polymorphism

It can be expected that practically all organic molecules that form stable crystals will tend to build different crystal structures (polymorphism). Differences in energies, and even more in free energies, of different polymorphs are quite small [6], which explains the coexistence of different polymorphs under the same thermodynamic conditions. Thus, even if a hypothetically 'perfect' method would yield a single crystal structure as the most stable one under given thermodynamic conditions, this does not necessarily mean that this structure is the one to be formed in a real crystallization experiment.

All that can be said is that the experimentally obtained crystal structure (provided the thermodynamic packing problem has been solved) has to be among the polymorphs with reasonable free energy.

1.2.2 Accuracy of the Energy Function

The empirical potentials used for computer simulations usually have an accuracy that is of the same order of magnitude as the energy differences between different polymorphs. In this context it seems somewhat unimportant that the influence of entropy during the search for the best structures is usually neglected, since its effect is anyway expected to be only marginal.

1.2.3 Global Search

The numerical problem of finding the global minimum of the energy function of a molecular crystal can become quite frightening. In the simplest case (space group $P1$, $Z = 1$) one has nine parameters to optimize: the lengths of the unit call vectors, the angles between them and three Euler angles to specify the orientation of the molecule treated as a rigid body. Assuming that a local energy minimum has been found, then it is very likely that changing one of the angles only slightly, a new local minimum in the vicinity of the new angular parameters will be found, if a change in the lengths of the cell vectors is allowed as well. It is easy to imagine that the energy function of a molecular crystal in general has an enormous number of different local minima and that therefore a search for the global minimum is by no means trivial, even for today's computers approaching the TFLOPS (i.e. 10^{12} floating-point operations per second) region.

2. METHODOLOGY

Despite the considerable difficulties involved in solving the molecular packing problem, a method [5] could be developed that generates a number of different polymorphs among which hopefully is the one(s) found experimentally (see Section 3.4.3 for the performance so far). In the following, this method will be explained in detail. Other methods will be discussed in Section 3.

2.1 History of the Present Method

In 1989, the Ciba-Geigy Company (Basel) planned to achieve the know-how necessary to solve actual problems of solid-state chemistry using methods of computational chemistry, that is, of molecular modeling. The scientific leader of the molecular modeling group (H. Karfunkel) had the task of managing this project, and it was realized that the solution of the packing problem was crucial for the solution of quite a long list of other problems.

At that time, the present author, after a PhD in quantum chemistry, obtained a postdoctoral position as a consultant for two years in the Ciba-Geigy molecular modeling group. The task was to search the literature for a method capable of solving the packing problem and, when it quickly became clear that such a method did not exist, to develop one. Relying on the basics of molecular mechanics, after virtually every obvious approach had been tried, the methodology described in this section was developed, and programming started, using the CHARMm [7] package through its user interface.

In 1990 the program was essentially finished and the first application to the crystal structure of hexamethylbenzene was tried. This structure was chosen because of its simplicity and because the compound is a pure hydrocarbon, since at that time it was questionable whether common force-fields would be accurate enough for molecules containing heteroatoms. Despite the fact that the packing was correct only in two dimensions, the first results were presented at the conference 'Elmau 3'.

In 1991, the present author considered the project successfully finished when the crystal structure of ethene was correctly predicted just by putting two molecules in the asymmetric unit and anticipating no symmetry at all; he then left Ciba-Geigy to continue to work on the development of quantum-chemical methods. Meanwhile, the Ciba-Geigy group then applied the software to molecules containing heteroatoms and, suprisingly, were successful. In 1992, the present author further improved the method together with F. Leusen, a PhD student, and in 1993 helped the Molecular Simulations Company, of Cambridge, England, to develop a commercial product which is now called 'Polymorph Predictor'.

2.2 General Considerations

2.2.1 Evaluation of the Energy Function

Force field. The potential energy of a molecular crystal, as a function of its structure, is readily evaluated when the interactions between all atoms are approximated by simple functions, i.e. like in an empirical force field (see Chapter 4). The parameters (with the exception of atomic charges, see next paragraph) can normally be transferred from one chemical system to another essentially without loss in accuracy.

Charges. Atomic charges are an exception because of the long-range nature of the electrostatic interaction. It turns out that for successful packing calculations, especially on polar (heteroatom-containing) molecules, it is advisable to fit the atom-centered charges to the electrostatic potential (ESP) of the isolated molecule. ESPs are readily obtained by quantum-chemical methods, e.g. *ab initio* Hartree–Fock (HF) [8], density functional, or, if CPU time is crucial, by semi-empirical HF, like e.g. MNDO [9, 10].

Polarization. In a crystal, the ESP arising from one molecule influences the electron density of the other molecules and vice versa. In general, this polarization is neglected. However, if the molecules contain polarizable atoms (e.g. sulfur) these may be adjacent in a real crystal, which in a computer simulation is rather unlikely to occur, since no attraction is generated when these atoms are permanently charged with the same sign. In the present method, charges are fixed; for the efficient inclusion of polarization in a force field, see e.g. [11] and references therein.

Ewald summation. Due to the above-mentioned long-range nature of the electrostatic interaction, care has to be taken in calculating this contribution. During the (somewhat time-consuming) optimization of a Monte Carlo (MC) trial structure (see Section 2.3.3), one is tempted to use a crystal fragment of only minimal size, i.e. only the interactions between neighboring molecules are accounted for when computing the non-bonding interaction. However, the contribution of these electrostatic interactions in the finally optimized MC trial structure may be largely in error when the molecules are polar, and almost totally wrong when additionally the unit cell possesses a net dipole moment. Therefore it is advisable (in the case of polar molecules, even mandatory) to use some form of convergence acceleration, e.g. the Ewald summation [12, 13] to obtain the energies of the final crystal structures that are eventually to be compared (see Section 2.4).

Beyond empirical potentials. In the present approach, it is not mandatory to use a parametric force field; in principle, and if the increase in CPU time can be afforded, one might use more sophisticated methods for treating solids, like, for example, a combination of quantum-mechanical and classical approaches [14] or perhaps even density functional theory.

2.2.2 Space-Group Symmetry

In general, neither the space group G nor the number ν of molecules in the asymmetric unit is known in advance. Therefore, when using the present method, the search for the global minimum of the potential-energy function of the crystal $E(\mathbf{x})$ with respect to the structural parameters $\mathbf{x}$, has to be performed for each relevant combination of G and ν. Alternatively, one could fix the space group to $P1$ (or to the simplest possible space group if the molecule is itself symmetric) and try various numbers of molecules in the unit cell. Fortunately, there are some statistical limitations to reduce the number of combinations that must be taken into consideration.

Recent work [15] has revealed that 92.71% of all organic crystal structures in the Cambridge Structural Database (CSD) [16] belong to one of only 18 space groups. These findings are in perfect agreement with theoretical considerations based on the dense packing of molecules [17]. Additionally, 91.7% of all crystal structures in the CSD have only one molecule in the asymmetric unit [18].

This implies that, if one picks an arbitrary compound from the CSD and scans the most common space groups, the number of molecules in the asymmetric unit being fixed as one, it is very probable that the correct combination of G and ν (see above) is covered. However, since the CSD cannot be expected to be representative for organic molecular crystals in general, in special cases care has to be taken, to check if the statistics described above are actually valid. For example, it is known that alcohols frequently crystallize with more than one molecule in the asymmetric unit [19].

Since the difficulty of the optimization problem can be expected to grow very fast with the number of degrees of freedom, and since, on the other hand, the search for the space group can hopefully be limited to a rather small subset of the 230 in total, for the general case, scanning the most probable space groups, rather than trying various values of ν in space group $P1$, is recommended.

2.3 Optimization by Smart Monte Carlo

2.3.1 Classification of the Structural Parameters

To uniquely define a crystal structure within a certain space group, it is sufficient to define three linearly independent vectors that span the unit cell, and the atomic coordinates in the asymmetric unit. In our method, the parameters necessary to define the crystal structure are (as usual) classified as (i) intramolecular (class A, class I in Ref. [5], that is, internal coordinates, such as bond lengths, bond angles, and dihedral angles; and (ii) intermolecular (classes B and C).

The intramolecular parameters x_{i_A} (class A) are kept fixed in the Monte Carlo simulation described below, that is, molecules are treated as rigid bodies. The proposed treatment of flexible molecules involves an additional methodology as explained in Section 2.3.4.

To make use of the special treatment described below, the intermolecular parameters are further divided into two classes. Parameters x_{i_B} of class B (class II in Ref. [5]) govern the spatial extension of the crystal's unit cell. These are the lengths of the unit cell vectors, the distances between the (arbitrary) first molecule and the others in the asymmetric unit, and, in some space groups, the distance from the first molecule of the asymmetric unit to the origin.

The remaining parameters x_{i_C} (class C, class III in Ref. [5]) are angles, that is, the potential-energy function of the crystal is a periodic function of these angles:

$$\forall(i_\mathrm{C})\exists(P_{i_\mathrm{C}} > 0)\colon E(x_{1_\mathrm{A}}, \ldots, x_{n_\mathrm{A}}, x_{1_\mathrm{B}}, \ldots, x_{n_\mathrm{B}}, x_{1_\mathrm{C}}, \ldots, x_{i_\mathrm{C}} + p_{i_\mathrm{C}}, \ldots, x_{n_\mathrm{C}}$$
$$= E(x_{1_\mathrm{A}}, \ldots, x_{n_\mathrm{A}}, x_{1_\mathrm{B}}, \ldots, x_{n_\mathrm{B}}, x_{1_\mathrm{C}}, \ldots, x_{i_\mathrm{C}}, \ldots, x_{n_\mathrm{C}}) \quad (1)$$

These parameters are the rigid-body rotations of the molecules in the asymmetric unit, the orientations of the vectors describing the locations of the molecules in the asymmetric unit, and the orientations of the lattice vectors **a**, **b** and **c**.

2.3.2 The Auxiliary Function C

The classification of the structural parameters in the previous section has been introduced to allow a special treatment, which is described in this section and which considerably simplifies the search for the global minimum of the energy function $E(\mathbf{x})$. It is a well-known fact that (with the exception of torsionally flexible molecules) intramolecular coordinates only change marginally when going from the isolated molecule to a molecular crystal [20, 21]. Therefore, as already mentioned in Section 2.3.1, class A parameters are held fixed during the search for the global energy minimum.

The parameters of class B are, in a sense, dependent on the parameters of the other classes, A and C: given a molecular conformation x_a and a set of values for the 'orientational' parameters $\mathbf{x}_C$, then the values of the parameters of class B may be defined by requiring that the resulting crystal be densely packed. We define an auxiliary function, $C(\mathbf{x}_C, \mathbf{x}_A)$, whose local minima of interest (see above) are to a very good approximation equal to corresponding local minima of $E(\mathbf{x}_A, \mathbf{x}_B, \mathbf{x}_C)$, the potential-energy function of the crystal within the current force-field approximation:

$$C(\mathbf{x}_C; \mathbf{x}_A) := \min_{\mathbf{x}_B} E(\mathbf{x}_B, \mathbf{x}_C; \mathbf{x}_A) \tag{2}$$

According to the previous argument, the values of the parameters of class A are held fixed throughout the simulation which is denoted by '$\ldots; \mathbf{x}_A$'. In general, one would expect more than one *local* minimum of E as a function of $\mathbf{x}_B$ only. Thus a bias must be prevented, e.g. by allowing for a certain 'randomness' of the search direction within the algorithm.

The number of independent structural parameters is thus vastly reduced. In the simplest case of space group $P1$ with one molecule in the asymmetric unit, only 6 independent structural parameters, namely 3 for the rigid body orientations and 3 for the orientations of the unit cell vectors are left. The lengths of the unit cell vectors depend on the previously mentioned parameters and have to be optimized according to Eq. (2). A detailed description of an algorithm capable of performing the evaluation of the C-function as defined by Eq. (2) is given in Ref. [5].

2.3.3 Optimization of the C-function Using Monte Carlo Simulated Annealing

Even when the number of structural parameters to be optimized has been largely reduced by defining the auxiliary function C, Eq. (2), finding its global minimum is still by no means trivial. To understand this point (recall the discussion of Section 1.2.3), consider a densely packed molecular crystal. Consider now a change in the i-th orientational parameter x_{i_C} by some amount, e.g. a rotation of the content of the asymmetric unit or a change in the orientations of the unit cell vectors, which is equivalent to shifting the layers in the crystal. According to the definition of C, Eq. (2), the parameters of class B now have to be optimized. Since the intermolecular interaction by definition is (with the exception of hydrogen bonds) generally weak, the new value of C will not differ much from the initial one, despite the fact that the $\mathbf{x}_B$ values may have undergone rather large changes. Furthermore, one can expect to find a rather large number of local minima when moving around the potential-energy hypersurface given as a function of one x_{i_C} only. Let N be an estimate for the number of local minima of C for each x_{i_C}; then, the total number of different local minima is N^{n_C}, n_C being at least 6 (as discussed in Section 2.3.2). Thus the total number of local minima may easily reach huge values.

A quite efficient optimization method, which has proved to be useful for 'difficult' cases, is so-called simulated annealing (SA) [22]. In SA, one treats the optimization problem as a (thermodynamical) random walk on an energy surface defined by the function whose global minimum is being searched for. The random walk is simulated by Monte Carlo using the Metropolis algorithm [23] where even a trial step that raises the energy ($\Delta E > 0$) is accepted with a non-zero probability equal to $\exp(-\Delta/kT)$. This provides the system with a drive to overcome energy barriers on the surface.

The trial displacements are computed by

$$x_i \rightarrow x_i + p_i \alpha \xi \tag{3}$$

where p_i are the periodicities defined in Eq. (1) of Section 2.3.1, α is the maximum allowed displacement in units of p_i ($0 < \alpha \leqslant 1$), and ξ is a random number uniformly distributed in the interval $(-\frac{1}{2}, \frac{1}{2})$.

The SA is started with $\alpha = 1$ and a temperature T_{init} high enough to easily overcome any barrier in the auxiliary function C. With $\alpha = 1$, there is at least the possibility that any crystal structure, also the 'correct' one, can be reached in just one step. Thus, the method is definitely ergodic. After each accepted step, α is halved and the temperature is slowly lowered by multiplying it by a fixed number. If the last step could not be accepted, α is doubled instead (if $\alpha < 1$) and the temperature remains unchanged. This procedure is continued until either the random walk is trapped ($\alpha < \alpha_{\min}$) or the crystal is 'frozen' ($T < T_{\min}$) (see Ref. [5] for details).

2.3.4 Suggested Treatment of Flexible Molecules

Finding the best conformers of a molecule containing many rotatable bonds alone can be a formidable task. Therefore, it is recommended that the conformational problem be separated from the packing problem, the former being supposed to have been solved before the latter.

Thus, initially one has to know the most important molecular conformers and compute a table whose entries are measurements of similarity between any given pair of conformers. Which molecular conformer actually is loaded during the generation of a new crystal trial structure in the SA, is decided by using a random number, with the current value of α (see the previous section) as well as the respective similarity to the last accepted conformer taken into account. At the beginning of the SA, where $\alpha = 1$, any conformer can be selected with an equal probability, but when α gets smaller, only a distinct subset can be used, such that a certain similarity to the last accepted conformer is ensured. At the end of the SA (where α is about zero) the random walk will be constrained to a single molecular conformer.

2.4 Data Reduction of Monte Carlo Results

The previously described Monte Carlo stage generates a trajectory file which contains the coordinates of the crystal structure that have been generated. The number of different structures is, in general, much too large for a more sophisticated treatment. However, there is much redundant information contained in the trajectory file, because, due to the Metropolis criterion [23], the MC structures cluster around local minima, hopefully containing the structures lowest in energy. The structures of each of these clusters are, of course, very similar to each other and thus, for further treatment, it is sufficient to choose a representative one from each cluster.

2.4.1 Cluster Analysis

This task is a standard problem of numerical mathematics; however, due to the quite large number N of different structures (of the order of several thousands) one has to make sure that the CPU time needed grows sufficiently slowly with increasing N. Comparing all possible pairs would involve $N(N-1)/2$ comparisons, which would be prohibitive for large N. Fortunately, one can use the energy values of the different structures as an auxiliary criterion.

The complete clustering would then involve the following steps:

(a) search the trajectory file for the structure with lowest energy;
(b) compare this structure with all the other structures on the file and incorporate into the cluster those that are close in energy and marking them for deletion in the trajectory file;
(c) repeat these steps until the energy of the representative structure becomes so high that it is unlikely that the particular structure is of importance.

The CPU time required for the clustering of, say, M distinct structures is proportional to N times M, and therefore the algorithm described above seems to be well suited for even very extensive MC runs.

2.4.2 Comparison of Two Structures

A straightforward criterion for comparing two different crystal structures would be the root-mean-square (RMS) distance between the atomic coordinates where, of course, the two unit cells should be transformed to the Niggli [24] reduced form. However, it frequently turns out that very similar structures may have entirely different reduced cells and also molecules that possess a symmetry axis of rotation may have undergone this symmetric rotation during the MC. The result is that, in such a case, the two very similar structures will not be recognized as such, unless the recognition procedure includes a symmetry check.

To solve this problem, the following two more sophisticated criteria for similarity of two crystal structures may be used.

Distances. Instead of comparing the coordinates of two different crystal structures, one might base the comparison on the interatomic distances in each of the structures. Such an algorithm would proceed as follows: (i) compute those distances between two atom types in each structure that are below a certain threshold; (ii) sort the distances and remove redundant values; (iii) compare the distances using a simple RMS measure (see above).

X-ray powder diagrams. Alternatively, one may compute the X-ray powder diagram of each structure as a 'fingerprint' and compare these diagrams (see e.g. Ref. [25] for details).

2.5 Final Refinement

The best crystal structures are further refined by a simultaneous optimization of *all* structural parameters ($\mathbf{x}_A$, $\mathbf{x}_B$ and $\mathbf{x}_C$) using a variable-metric algorithm (see e.g. Ref. [26]). One might think of using molecular dynamics to check the stability of the polymorphs. Free energies could also be computed, at least including vibrational entropies in a harmonic approximation. Finally, unique crystal structures are obtained by performing a last cluster analysis.

2.6 Summary

Applying the packing algorithm described above to predict possible molecular crystal structures is straightforward within the framework of common molecular modeling software. The generation of a set of possible crystal structures using the present method involves the following steps:

(a) preparation of the molecular structure(s), including all the most relevant molecular conformations in the case of flexible molecules (see Section 2.3.4);
(b) calculation of atom-centered charges by a fit to the electrostatic potential of the molecule(s) as described in Section 2.2.1;
(c) for each relevant space group (about 20 usually have to be considered) the following three steps have to be performed: generation of the trajectory using simulated annealing with a special constrained Monte Carlo (see Section 2.3.3); cluster analysis of the result (see Section 2.4.1); and refinement of the best structures by optimization of all structural parameters (see Section 2.5);
(d) final cluster analysis to yield unique structures and evaluation of the results.

Some examples of the applications of the above-described procedure will be given in Section 3.4.3.

3. PUBLISHED METHODS AND THEIR APPLICATION

3.1 Restricted Packing Problem

3.1.1 Anticipation of the Unit Cell

If the unit cell parameters are known, it is often quite simple to find a possible arrangement of the molecules in the unit cell. Kitaigorodski [27, 17] has demonstrated this using a mechanical apparatus (called a 'structure finder'). This approach is readily extended to force-field methods and provided the structure of the molecules in the crystal is known in advance, one can expect to obtain the correct structure rather easily (see e.g. Ref. [28] for a recent application).

3.1.2 Packing in One Dimension

Perlstein [29] optimized one-dimensional chains of molecules obtained by applying the most common symmetry elements (i.e. the inversion center, the screw axis and the glide plane) and then using Monte Carlo simulated annealing. Of 60 crystal structures of molecules containing C, H, O, N, F, Cl, Br, I, S, having no hydrogen bonds and only one molecule in the asymmetric unit, in 53 cases, a local energy minimum for the string could be obtained that corresponds to a structure close to the one found experimentally, its energy being 0.1–5.7 kcal/mol above the global minimum. A method using these one-dimensional chains to build the three-dimensional crystal structures is proposed.

3.1.3 Packing in Two Dimensions

Zorkii and Porai-Koshits [30, 31] proposed a graphical method to determine the closest-packed layer of molecules. The method has successfully been applied to crystals of nickel dimethylgloxime [32] and copper salicylalmethyiminate [33]. In both cases, the experimentally found structures contained layers which were close-packed, but did not correspond to the closest packing.

3.2 *Ab initio* Packing of Molecular Clusters

3.2.1 Packing of Infinite Chains

Zorkii's method of 'symmetry of the potential functions' [34] was applied to the packing of infinite chains occurring in the crystals of palladium chloride [35] and beryllium chloride [36]. Again, dense packings were calculated and it was found that in the case of $PdCl_2$ the experimental structure was among the densest obtained, whereas in the case of $BeCl_2$ the experimental structure was actually close to the most closely packed one, as determined from the calculations.

3.2.2 Packing of Three-Dimensional Clusters

Gavezzotti [4] proposed a method based on using the most common symmetry elements occuring in molecular crystals (see Section 3.1.2) to build clusters consisting of 2–4 molecules, which, in turn, are translated to form molecular crystals. The energies of the clusters and of the molecular crystals were computed using empirical potentials. Additionally, statistical correlations between molecular and crystal properties were used to guide the search. During the generation process, some decisions had to be made by the user. The method has been applied to 18 hydrocarbons of known crystal structure and in 10 cases the correct structure could be found. Additionally, predictions concerning a hydrocarbon whose crystal structure was unknown, were made (see Chapter 1 for more details and developments of the method).

3.3 Static *ab initio* Approaches within the Crystal Environment

3.3.1 Artificial Initial Structures

Shoda et al. [37] were able to obtain a low- and a high-pressure modification of benzene crystals by optimizing artificial initial structures using the method of conjugated gradients within a force-field approximation. These polymorphs turned out to be close to experimentally found crystal structures. The stability of the polymorphs was confirmed using molecular dynamics. A very similar approach was independently used by Gibson and Scheraga [38, 39] (see Ref. [39] also for previous investigations on the crystal structure of benzene), who optimized 60 artificial crystal structures of benzene, obtaining structures that came close to experimentally observed polymorphs.

3.3.2 Random Initial Structures

Schmidt [40] proposed the optimization of randomly chosen crystal structures. During the packing procedure, the molecules were treated as rigid bodies whose pairwise interaction was obtained using empirical potentials. The search procedure was assisted by manual intervention. The method was applied to various hydrocarbons, silicon-containing molecules, and organo-metallics (25 compounds in total), and in all cases, the best structures found were close to the experimental one.

Williams [41] proposed the 'off-ridge eigenvector minimization' (OREM) which performs identically to Newton–Raphson (NR) or steepest descent (SD), provided that, during the search, the Hessian matrix stays positive definite. However, this condition is frequently violated in molecular mechanics optimizations, due to saddle points on the potential-energy surfaces. NR will rather be *attracted* to these saddle points, since it is basically an algorithm to *stationarize* functions rather than explicitly *minimize* them. If negative eigenvalues appear in the Hessian matrix, the OREM will make a step in exactly the direction of the eigenvector corresponding to the smallest negative eigenvalue, thus pointing (in the quadratic approximation) to the region of energy decrease. This method has very recently [42] been used to correctly reproduce the crystal structures of benzene at low pressure ($Pbca$, $Z = 4$), of benzene at high pressure ($P2_1/c$, $Z = 2$) and of urea ($P\bar{4}2_1m$, a space group with extremely low occurrence, $Z = 2$). As a starting model, a large body-centered ($Z = 2$) or face-centered ($Z = 4$) cubic unit cell was assumed, the molecules were randomly oriented in space, and then OREM was used to minimize the energy.

3.3.3 Grid Search

Recently, van Eijck et al. [43] in a *tour de force* attempted the crystal-structure prediction of six hexapyranoses for which experimental structures are available and that crystallize with one molecule in the asymmetric unit (α-D-galactose, α-D-glucose, α-D-talose, β-D-allose, β-D-galactose and β-D-glucose; see Chapter 4 for more details). Over 8 000 000 trial structures were generated, and then reduced to some 1000 polymorphs, being found to lie in the energy range of 10 kcal/mol. In all cases, the experimental structure was close to one of the generated structures, and in two cases, these structures also had the best energy.

3.3.4 Initial Structures Built from Known Coordination Patterns

Holden et al. [44] analyzed the crystal environments of organic molecules containing nitrogen, oxygen and fluorine in 242 crystal structures, to determine the common coordination sphere patterns. The crystal structures were then generated by trying to apply all appropriate coordination patterns around a central molecule. Finally, the densest packings were optimized in an empirical force field. The method has been applied to nine molecules containing nitro groups and in all cases the predicted structures were very close to the structures obtained by optimization of the experimentally observed ones.

3.4 Dynamic *ab initio* Approaches within the Crystal Environment

3.4.1 Molecular Dynamics

Shoda et al. [45] recently published a molecular dynamics (MD) study on benzene crystals. Starting from an artificial cubic crystal with $a = 10$ Å and four benzene molecules being stacked in parallel, MD at constant volume (the cell shape was fixed at this stage) and temperature ($T = 500$ K) was used to generate 20 initial structures. In the second step, the energy was minimized by the method of conjugated gradients, with the cell shapes being allowed to vary. In the third step, the crystal structure was subjected to MD at constant stress and enthalpy (again at 500 K). The last two steps (called 'pseudo-annealing') were performed ten times. Finally, in a fourth step, the relative pressure stability of the structures was investigated by a minimization of the enthalpy at pressures from 0 to 4 GPa in intervals of 0.5 GPa. The most stable structures obtained by the simulation at pressures up to 1.4 GPa, and above 1.4 GPa, were found to be close to the experimentally observed phases I (low pressure) and III (high pressure).

A very similar approach has independently been proposed by Hirano et al. [46]. Assuming that there are 1, 2, 4, or 8 molecules in the unit cell–it is known [47] that 84.7% of all homomolecular crystals belong to this set–8 molecules are placed so as to form $2 \times 2 \times 2$ blocks of a simple cubic lattice, the orientation of each molecule being set randomly. Then an MD run at constant temperature and constant pressure is performed, where the unit cell may change ('shaking'). The structures lowest in energy encountered in the MD are then clustered and optimized ('quenching'). This so-called molecular dynamics for crystal packing (MDCP) method has so far been applied to carbon dioxide [46, 48], where the experimental structure was equal to the best predicted; to benzene [46], where a run at 1 bar pressure yielded a structure lowest in energy that was close to the experimentally found low-pressure polymorph, while the structure of the experimentally found high-pressure polymorph was found to be close to the structure with the fourth best energy; to pyrimidine [46], where the global energy minimum could not be found, since the best predicted structure was higher in energy than the one obtained by optimizing the experimentally found one; and to 1, 2-dimethoxyethane [49], where the most probable crystal structure contained a conformer that is consistent with the results of infrared and Raman spectroscopy on the solid.

3.4.2 Off-ridge Eigenvector Minimization with Simulated Annealing

The 'off-ridge eigenvector minimization' (OREM) described in Section 3.3.2 has been combined with a simulated annealing schedule (OREMWA) [41], which interrupts the minimization when the last improvement in energy is smaller than kT (compare Section 2.3.3). If this happens, the temperature is lowered and a single random move (compare Eq. (3) of Section 2.3.3) limited to some range, is made. The range parameter (α in Eq. (3)) is decreased and the OREM is continued. The optimization is stopped when no significant further progress can be made.

This method has very recently been applied to the prediction of the crystal structure of *m*-nitroaniline [50]. One molecule was placed in a cubic unit cell ($a = 12$ Å) and oriented randomly in space. The experimentally observed symmetry ($Pbc2_1$) was imposed. Several runs of OREMWA with different starting structures were performed. The best structure obtained was close to the one found experimentally.

3.4.3 Smart Monte Carlo Simulated Annealing

The method described in Section 2 has been applied to the crystal-structure prediction of azo-bis (isobutironitrile) (CSD refcode AZIBYD) [51]; cyclo-bis (dehydro-alanyl), DAWSUP [51, 52, 53]; cyclo-L-alanyl-L-alanyl, LCDMPP [51, 53]; 4,8-dimethoxy-3,7-diazatricyclo(4.2.2.2-2,5-)dodeca-3,7,9,11-tetraene (BAWHOW) [51]; ethene and hexamethylbenzene [5]; isoiridomyrmecin [51, 52]; and a pyrrolo-pyrrol derivative called 'pigment red' [25]. In all these cases, the application of the present method yielded a number of probable structures that contained the experimentally found one(s) within the accuracy allowed by the force field. Provided that the force field used is sufficiently accurate – especially, that the electrostatic contribution is treated with care (see Section 2.2.1 for details) – until now there has been no case where an experimentally observed crystal structure could not be found among the ones generated by the present method.

Additionally, based on the successful generation of known crystal structures of LCDMPP and DAWSUP, the existence of two previously unknown structures could be claimed [54, 53].

4. CLOSING REMARKS

At present, the *ab initio* prediction of possible molecular crystal structures is still controversial. However, comparing the present situation with the state of the art just a few years ago, it is evident that enormous progress has been made. Indeed, using the computational methodology described in this paper, the generated information concerning possible crystal structures of molecular compounds, under favorable circumstances, may lead to a considerable lowering of the costs and effort involved in the research and development of a chemical product to be used in the solid state. Clearly, the methods presented here do not yet satisfy all needs, but it is the opinion of the present author that most of the main obstacles have been removed, so this exciting subject is becoming more and more popular among computational scientists. Keeping this in mind, one may find appropriate the closing remarks of Maddox's article [1]: 'In applying calculations to these needs, a demonstration of success can rank, pyschologically, with the example set by those who first climbed Everest'.

5. ACKNOWLEDGEMENTS

Thanks are due to Prof. H. Deuling and Prof. D. Kolb for helpful comments on the manuscript, as well as to Prof. D. E. Williams for a preprint of his manuscript '*Ab initio* packing analysis'.

6. REFERENCES

1. J. Maddox, *Nature*, **335**, 201 (1988).
2. F. C. Hawthorne, *Nature*, **345**, 297 (1990).
3. P. J. Fagan and M. D. Ward, *Scientific American*, **267**(1), 28 (1992).
4. A. Gavezzotti, *J. Am. Chem. Soc.*, **113**, 4622 (1991).
5. R. J. Gdanitz, *Chem. Phys. Lett.*, **190**, 391 (1992).
6. A. Gavezzotti, *Acc. Chem. Res.*, **27**, 309 (1994).
7. B. R. Brooks, R. E. Bruccoleri, B. D. Olafson, D. J. States, S. Swaminathan and M. Karplus, *J. Comp. Chem.*, **4**, 187 (1983).
8. S. R. Cox and D. E. Williams, *J. Comp. Chem.*, **2**, 304 (1988).
9. B. H. Besler, K. M. Merz and P. A. Kollman, *J. Comp. Chem.*, **11**, 431 (1990).
10. K. M. Merz and B. H. Besler, *QCPE Bulletin*, **10**, 15 (1990).
11. T. P. Straatsma and J. A. McCammon, *Chem. Phys. Lett.*, **167**, 252 (1990).
12. P. P. Ewald, *Ann. Phys. (Leipzig)*, **64**, 253 (1921).
13. D. E. Williams, *Acta Cryst.*, **A27**, 452 (1971).
14. A. Warshel and M. Levitt, *J. Mol. Biol.*, **103**, 227 (1976).
15. W. H. Baur and D. Kassner, *Acta Cryst.*, **B48**, 356 (1992).
16. F. H. Allen, S. Bellard, M. D. Brice, B. A. Cartwright, A. Doubleday, H. Higgs, T. Hummelink, B. G. Hummelink-Peters, O. Kennard, W. D. S. Motherwell, J. R. Rodgers and D. G. Watson, *Acta Cryst.*, **B35**, 2331 (1979).
17. A. I. Kitaigorodski, *Organic Chemical Crystallography*, Consultants Bureau, New York, 1961.
18. N. Padmaja, S. Ramakumar and M. A. Viswamitra, *Acta Cryst.*, **A46**, 725 (1990).
19. A. Gavezzotti and G. Filippini, *J. Phys. Chem.*, **98**, 4831 (1994).
20. A. Warshel, E. Huler, D. Rabinovich and Z. Shakked, *J. Mol. Struc.*, **23**, 175 (1974).
21. A. Warshel, in *Semiempirical Methods of Electronic Structure Calculation*, edited by G. A. Segal, Part A., Plenum Press, New York, 1977, p. 133.
22. S. Kirkpatrick, C. D. Gelatt, Jr. and M. P. Vecchi, *Science*, **220**, 671 (1983).
23. N. Metropolis, A. W. Rosenbluth, M. N. Rosenbluth, A. H. Teller and E. Teller, *J. Chem. Phys.*, **21**, 1087 (1953).
24. P. Niggli, *Kristallographische und strukturtheoretische Grundbegriffe. Handbuch der Experimentalphysik, Part 1*, Vol. 7, Akademische Verlagsgesellschaft, Leipzig, (1928).
25. H. R. Karfunkel, B. Rohde, F. J. J. Leusen, R. J. Gdanitz and G. Rihs, *J. Comp. Chem.*, **14**, 1125 (1993).
26. R. Fletcher, *Comp. J.*, **13**, 317 (1970).
27. A. I. Kitaigorodski, *Izw. AN SSSR. Otd. khim.* n. 6, 587 (1946).
28. W. Linert and F. Renz, *J. Chem. Inf. Comp. Sci.*, **33**, 776 (1993).
29. J. Perlstein, *J. Am. Chem. Soc.*, **116**, 455 (1994).
30. P. M. Zorkii and M. A. Porai-Koshits, *Sov. Phys. Cryst.*, **6**, 529 (1962).
31. P. M. Zorkii and M. A. Porai-Koshits, *Sov. Phys. Cryst.*, **12**, 863 (1968).
32. P. M. Zorkii and M. A. Porai-Koshits and V. S. Psalidas, *Zh. Strukt. Khim.*, **7**, 577 (1966).
33. P. M. Zorkii, M. A. Porai-Koshits and S. N. Sveshnikova, *Zh. Strukt. Khim.*, **7**, 752 (1966)
34. P. M. Zorkii, *Kristallografiya*, **13**, 26 (1968).
35. P. M. Zorkii and M. A. Porai-Koshits and L. M. Prezman and G. P. Murav'eva, *Zh. Strukt. Khim.*, **10**, 633 (1969).
36. L. M. Prezman and P. M. Zorkii, *Zh. Strukt. Khim.*, **12**, 1044 (1971).
37. T. Shoda, K. Yamahara, K. Okazaki and D. E. Williams, *J. Mol. Struc.*, **313**, 321 (1994).

38. K. D. Gibson and H. A. Scheraga, *J. Phys. Chem.*, **99**, 3752 (1995).
39. K. D. Gibson and H. A. Scheraga, *J. Phys. Chem.*, **99**, 3765 (1995).
40. M. U. Schmidt, *Berichte aus der Chemie*. Verlag Shaker, Aachen, 1995. Thesis RWTH Aachen.
41. D. E. Williams, *Chem. Phys. Lett.*, **192**, 538 (1992).
42. D. E. Williams, *Acta Cryst.*, **A53**, 326 (1996).
43. B. P. V. Eijck, W. T. M. Mooij and J. Kroon, *Acta Cryst.*, **B51**, 99 (1995).
44. J. R. Holden, Z. Du and H. L. Ammon, *J. Comp. Chem.*, **14**, 422 (1993).
45. T. Shoda, K. Yamahara and D. E. Williams, *J. Mol. Struc.*, **333**, 267 (1995).
46. N. Tajima, T. Tanaka, T. Arikawa, T. Sakurai, S. Teramae and T. Hirano, *Bull. Chem. Soc. Jpn.*, **68**, 519 (1995).
47. V. K. Belsky and P. M. Zorkii, *Acta Cryst.*, **A33**, 1004 (1977).
48. T. Hirano, S. Tsuzuki, K. Tanabe and N. Tajima, *Chem. Lett.*, 1073 (1995).
49. T. Arikawa, N. Tajima, S. Tsuzuki, K. Tanabe and T. Hirano, *J. Mol. Struc.*, **339**, 115 (1995).
50. T. Shoda and D. E. Williams, *J. Mol. Struc.*, **357**, 1 (1995).
51. H. R. Karfunkel and R. J. Gdanitz, *J. Comp. Chem.*, **13**, 1171 (1992).
52. H. R. Karfunkel and F. J. J. Leusen, *Speedup*, **6**, 43 (1992).
53. H. R. Karfunkel, F. J. J. Leusen and R. J. Gdanitz, *J. Comp. Aided Material Design*, **1**, 177 (1993).
54. R. J. Gdanitz, F. J. J. Leusen and H. R. Karfunkel, *J. Mol. Graphics*, **11**, 275 (1993).

Chapter 7

The Crystal Habit of Molecular Materials: A Structural Perspective

Graham Clydesdale, Kevin J. Roberts and Elaine M. Walker
Centre for Molecular and Interface Engineering, Department of Mechanical and Chemical Engineering, Heriot-Watt University, Edinburgh

1. INTRODUCTION

Crystallisation is an important energy-efficient process for the purification and separation of bulk and fine chemicals as well as being a significant unit operation in the manufacture of a wide range of speciality chemicals and intermediates such as agrochemicals, pharmaceuticals, dyestuffs and monomer units for high-performance polymers. Predicting, controlling and manipulating crystal shape produced in such processes can have important ramifications on the overall manufacturing process. For example, an inadequately defined morphology may have a detrimental impact on a range of particulate properties such as:

(a) reduced separation efficiency, due to difficulties in filtration;
(b) creation of hazardous dusts, due to attrition and fragmentation during particle flow;
(c) caking, due to a poorly defined packing density of the product;
(d) variable dissolution rate, due to variable surface area.

The optimisation of crystal shape in industrial processes has, in the past,

Theoretical Aspects and Computer Modeling Edited by A. Gavezzotti

tended to be based on empirical approaches which are, by their nature, system-specific and thus give rather limited scope for process development and enhancement. Such a perspective has evolved a need for an improved understanding of the underlying molecular processes taking place during crystallisation. This has led, in turn, to a significant interdisciplinary effort, involving chemists, computational modellers, crystallographers and process engineers, being devoted towards understanding and predicting the external crystal shape in terms of the internal solid-state structure [1–4]. These investigations have been applied not only to systems of academic interest but also to a number of important industrial problems [5], such as studying the structural role played by crystal habit-modifying additives [6–9] in order to engineer a desired crystal shape.

In this chapter we present an overview of the science underpinning the prediction of crystal shape together with the computational methodologies adopted. The applications are illustrated through a number of predictive examples. All of the calculations were performed using HABIT95 [10] (see Appendix) a computer program developed within this research group.

2. UNDERLYING CRYSTAL GROWTH THEORY

Crystallisation from the liquid phase is a supersaturation-driven process that can be subdivided into two distinct phases: nucleation and growth. In the nucleation stage, supersaturation causes the formation and agglomeration of molecular solute clusters in solution. These clusters are maintained in a dynamic equilibrium that is dictated by the balance between the volume and surface contributions to their excess free energy. At sufficient cluster sizes, the bulk term dominates over that due to the surface and a viable crystalline particle is formed. This nucleation stage is genuinely a 3-D process in contrast to the growth process that follows in which each stable cluster or grain develops by a co-operative series of growth events taking place on a number of structurally distinct crystal-growth interfaces (see Figure 7.1). A 3-D crystallite grain is in reality just the addition of a number of 2-D events that occur on individual crystal surfaces which are usually aligned on low-index Miller planes (*hkl*).

Over the past 40 years our understanding of growth-interface kinetics has developed significantly. The foundation for research in this area was laid by the seminal work of Burton, Cabrera and Frank (BCF) [11] which has been the basis for the development of crystal growth theory in the years since. The main advances in the theory have been in our understanding and definition of the three mechanistic regimes of crystal growth (see Figure 7.2) that define the relationship between the surface growth rate (R_{growth}) and the supersaturation (S) of the crystallising system:

(a) at low supersaturations, the BCF mechanism requires screw dislocations to

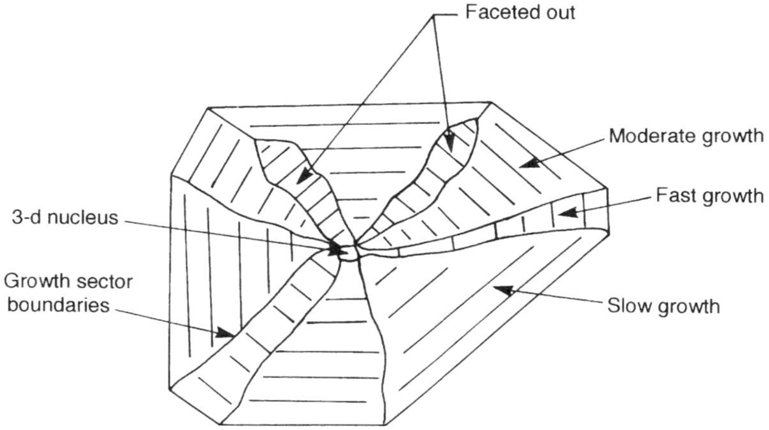

Figure 7.1 Schematic showing a cut made through a 3-D crystal, revealing its sectorial nature, with crystal growth simultaneously proceeding in a number of growth interfaces

provide crystal-surface attachment sites, enabling growth at the crystal/liquid interface to proceed with a parabolic rate dependence, i.e. $R_{growth} \propto S^2$;

(b) at moderate supersaturations surface integration can take place by 2-D surface nucleation via a birth and spread (B&S) mechanism (see e.g. [12]); here the rate law becomes $R_{growth} \propto S^{5/6} \exp(S)$;

(c) at high supersaturations a rough interface growth (RIG) model occurs. Interface (kinetic) roughening (e.g. [13]) takes place and growth can proceed without the need to form any well-defined surface layers at the interface. A linear rate dependence ($R_{growth} \propto S$) occurs, where the growth interface is inherently unstable.

Due to the different surface chemistries of the various crystal faces, each crystallographically unique face may grow by any one of the three interface kinetic processes. This variation in surface chemistry and hence surface binding with crystal orientation means that the growth and development of a crystal is likely to be anisotropic. In addition, depending on the relative growth rates of the crystal faces in three dimensions, the particle morphology can be highly variable. Thus the size and shape of a single particulate grain is governed by the number of crystallographically independent faces and by which of the three mechanisms the faces grow. The implications of morphological control in industrial crystallisation technology mean that the surface processes require the particle morphology to be maintained, such that none of the crystal surfaces should grow under an RIG mechanism, if the resulting particle is to avoid becoming excessively contaminated with impurities.

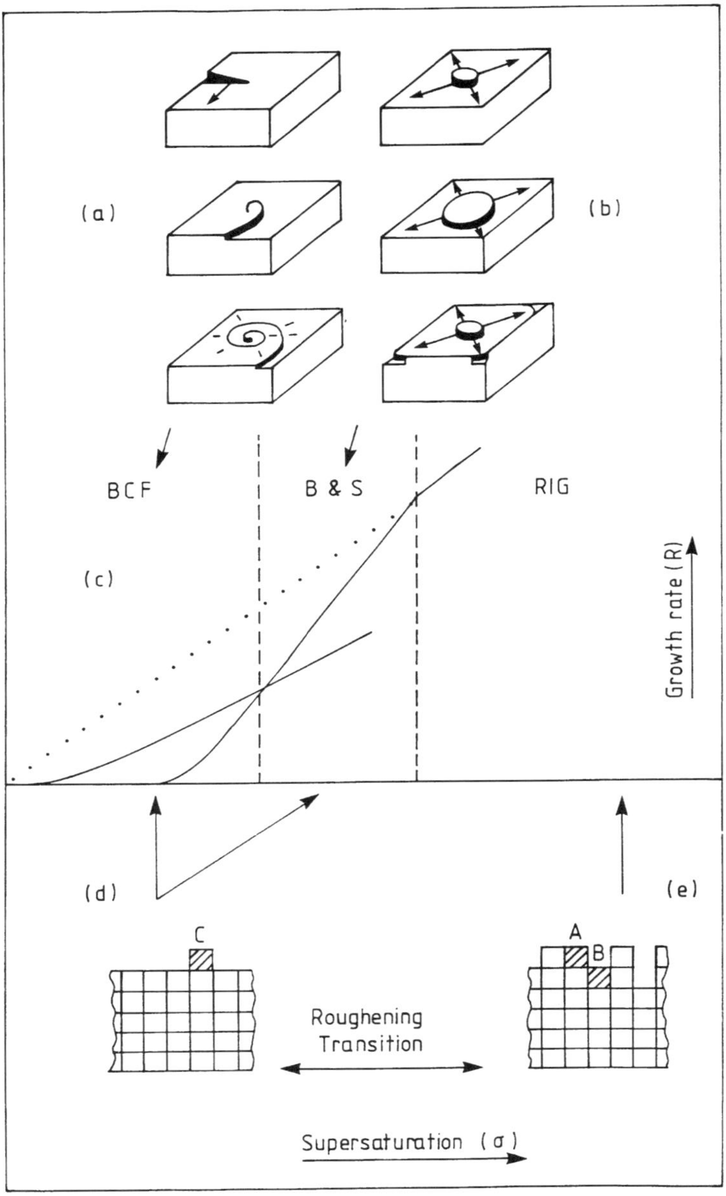

Figure 7.2 Schematic illustrating the interrelationship between kinetic mechanisms and supersaturation: (a) the dislocation-controlled BCF mechanism; (b) two-dimensional surface nucleation by birth and spread mechanism; (c) the relation between mechanisms and supersaturation; (d), (e) rough-interface growth (RIG) showing the formation of kink (A) and step (B) sites after transformation from a singular (C) to an atomically rough interface

In this chapter, we concentrate on predictions of crystal morphology for surfaces that have grown under a B&S mechanism. We also restrict ourselves to the specific case of the crystal growth of organic solids.

3. INTERMOLECULAR INTERACTIONS AND SOLID-STATE STRUCTURE

Organic molecules can be considered as impenetrable units whose shape and volume are determined by the molecular conformation and the van der Waals radii of the constituent atoms [14]. In turn, this is determined by both internal factors, such as the nature of the interatomic (covalent) bonds, and effects imposed by the local environment around the molecule. Due to the exclusive nature of the molecule, no neighbouring units may penetrate within the atomic radii of any atom and the molecules remain mutually exclusive within the lattice. As organic molecules are often bulky and awkwardly shaped, the result is crystal systems of low symmetry, as revealed by the analysis of the Cambridge Structural Database [15] presented in Figure 7.3. These low symmetries are determined by translational symmetry elements in which the molecules sit in the voids created by the packing of the surrounding molecular units (see Chapter 1). As these materials are essentially loosely packed, the nature of the attractions that hold the molecules together in the lattice needs to be considered if the crystallisation process is to be modelled on a structural basis. The dominant intermolecular forces are: van der Waals attractions, hydrogen-bonding, electrostatic interactions, specific directional interactions (see Chapter 2). The sum total of all the intermolecular interactions in a crystal structure is expressed as the lattice energy of a crystal (E_{cr}) and reasonable approximations to this, via the atom–atom potential method (for example, see [16–22]) have been developed.

Calculation of crystal lattice energies can give information on both inter- and intramolecular interactions, as well as interatomic interactions. The lattice energy of a crystal can be calculated from a 3-D model of the crystal structure by summing the interactions between a central molecule and all the surrounding molecules out to a distance such that the intermolecular interaction becomes negligible. The atom–atom method considers each interaction to be the sum of the constituent atom–atom interactions. If there are n atoms in the central molecule and n' atoms in each of the N surrounding molecules then one gets:

$$E_{cr} = \frac{1}{2}\sum_{k=1}^{N}\sum_{i=1}^{n}\sum_{j=1}^{n} V_{kij} \tag{1}$$

where V_{kij} is the interaction energy between atom i in the central molecule and atom j in the kth surrounding molecule. Since the interactions are between two bodies, the factor of $\frac{1}{2}$ arises, to avoid double counting. The interaction V_{kij}

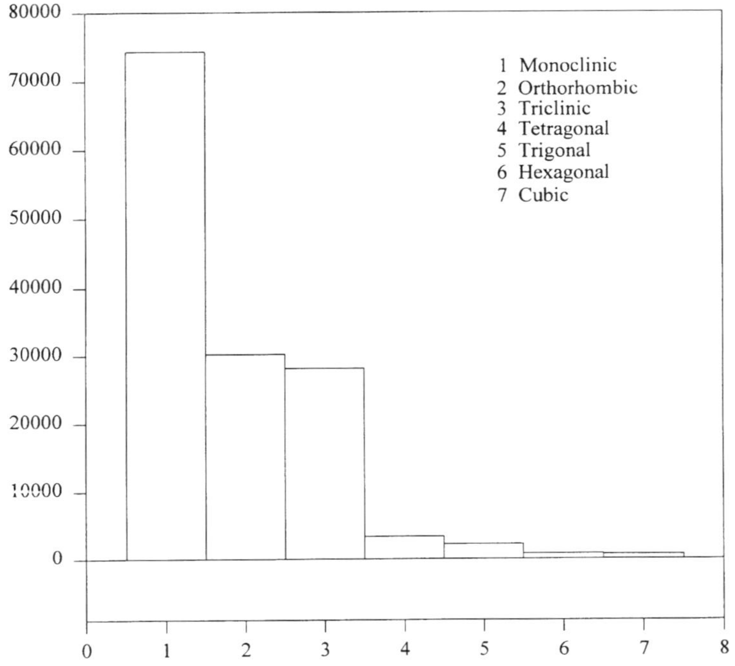

Figure 7.3 Graph showing the number of organic molecules within the Cambridge structural database, and the proportion of each crystal system

between any of the two non-bonded atoms i and j can be described by a number of intermolecular potential functions (see also Chapter 3, Appendix). Equation (2) gives the Lennard-Jones (6-12) interatomic function combined with a simple coulombic potential:

$$V_{ij} = -A/r^6 + B/r^{12} + q_i q_j/r \tag{2}$$

where A and B are empirical atom–atom parameters obtained from experimental and theoretical work and are dependent on the atom types interacting, q_i and q_j are the fractional charges placed on atoms i and j, and r is the interatomic distance. In some cases, where hydrogen bonds are important, other forms such as a 10-12 potential of similar structure to Eq. (2) can be used.

Another potential form which can be used is the Buckingham (6-exp) function of the form (including the coulombic contribution):

$$V_{ij} = -A/r^6 + B\exp(-Cr) + q_i q_j/r \tag{3}$$

where the parameters are defined in a similar manner to those given in Eq. (2)

with C as an additional atom–atom constant. The atom-pair values for the A and B parameters may be deduced using the combination rule:

$$A_{ij} = (A_{ii}A_{jj})^{1/2} \tag{4}$$

whereas for the C parameter in the Buckingham potential a geometric mean is used:

$$C_{ij} = (C_{ii} + C_{jj})/2 \tag{5}$$

The three terms in Eqs (2) and (3) represent the London dispersions (i.e. attractive energy), the repulsive energy and the Coulomb interactions respectively. Partial charges can be calculated, for example, using the semi-empirical quantum-chemistry methods (e.g. MNDO [23]) assuming no overall charge on the molecule.

The suitability of intermolecular potentials can be demonstrated by the comparison of the calculated lattice energy with the experimental lattice energy (V_{exp}); some examples are shown in Table 7.1. V_{exp} is estimated (see Chapter 3, Appendix) from sublimation enthalpy (ΔH_{sub}) data via Eq. (6):

$$V_{\text{exp}} = -\Delta H_{\text{sub}} - 2RT \tag{6}$$

where $2RT$ represents a correction factor for the difference between the gas-phase enthalpy and the vibrational contribution to the crystal enthalpy [21] (see also Chapter 3, Section 2). The results in Table 7.1 show that it is possible to carry out calculations on a range of materials with some confidence.

3.1 Terephthalic Acid [24]

Terethphalic acid (TA) has two polymorphs (α and β) both having space group $P\bar{1}$ with unit cell dimensions $a = 7.73$ Å, $b = 6.443$ Å, $c = 3.749$ Å, $\alpha = 92.75°$, $\beta = 109.15°$, $\gamma = 95.95°$ and $a = 9.540$ Å, $b = 5.340$ Å, $c = 5.020$ Å, $\alpha = 86.95°$, $\beta = 134.65°$, $\gamma = 104.90°$ respectively [25]. A numbered molecular

Table 7.1 Examples of calculated and experimental lattice energies

Material	Lattice energy (kcal/mol)			
	Calculated	Ref.	Experimental	Ref.
Adipic acid	−33.8	[17]	−32.1	[17]
Benzoic acid	−20.4	[17]	−23.0	[65]
Benzil	−25.9	[21]	−24.7	[66]
Trinitrotoluene	−25.1	[18]	−24.4	[65]
Urea	−22.7	[21]	−22.2	[65]

structure is shown below (assuming a centrosymmetric molecule):

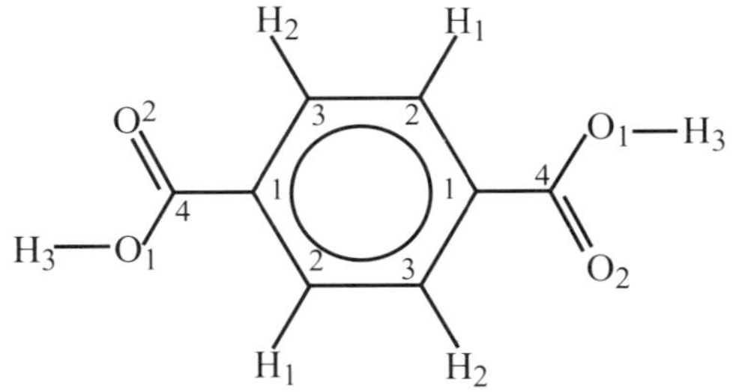

Examination of the crystal packing in each polymorph has shown the α-form to be characterised by hydrogen-bonding along the (110) plane at an O...H distance of 1.66 Å, with closest intermolecular C—H...O distance of 2.99 Å. The β-form, however, comprises hydrogen-bonded layers along the crystallographic a-axis at a distance of 1.66 Å and closest intermolecular C—H...O distance of 4.36 Å. Both crystal patterns have intramolecular C—H...O distances of 2.49 Å. Lattice energies of −37.19 and −35.29 kcal/mol were determined [20] for the α and β forms respectively, reflecting the experimental lattice energy of −34.4 kcal/mol [26].

The relative stability of these two forms, in terms of qualifying which polymorph is the more stable at room T, has been the subject of much interest [27–31]. Examination of the atom–atom and intermolecular interactions show that the carboxylic oxygen (O_2) and hydrogen (H_6) atoms contribute most to the overall lattice energy (see Table 7.2) and that the strongest intermolecular interactions are those along the direction of the respective hydrogen-bonded chains. Table 7.3 lists the direction of the strongest interaction in each of the polymorphs, along with the constituent atom–atom interactions that contribute

Table 7.2 Terephthalic acid crystal: atomic contributions to the lattice energy (%)

Atom	Atomic contribution to lattice energy (%) α polymorph	β polymorph
C_1	3.96	4.38
C_2	4.36	4.26
C_3	4.16	4.51
C_4	3.17	2.39
O_1	2.03	2.00
O_2	17.36	18.30
H_1	2.12	1.41
H_2	1.56	1.66
H_3	11.27	11.08

Table 7.3 Intermolecular and atom–atom interactions in terephthalic acid 'strongest bond'

Polymorph	UVW	Distance (Å)	Energy (kcal/mol)	Atom 1	Atom 2	Distance (Å)	Energy (kcal/mol)
α	110	9.54	−7.67	O_2	H_3	1.66	−7.82
				O_2	C_4	3.36	−3.94
				O_1	C_4	3.43	−2.44
				O_1	H_3	2.87	−1.82
β	100	9.54	−7.18	O_2	H_3	1.66	−7.80
				O_2	C_4	3.41	−3.88
				O_1	C_4	3.39	−2.47
				O_1	H_3	2.75	−1.91

to the energy. This is consistent with the morphologies of TA crystals, which comprise needles, with elongation along the hydrogen-bond direction, as a result of fast crystal growth.

4. PREDICTING CRYSTAL MORPHOLOGY

An attempt to quantify the energetics of the crystal growth process was considered by Gibbs in 1875 [32] and Wulff in 1901 [33] who suggested that the equilibrium form of a crystal would be such as to minimise the total surface energy for a given sample volume. Thus it became accepted that the 'as-grown' crystal would be bounded by a 3-D network of faces, each of which would be located at a distance, with respect to the growth nucleation centre, proportional to its surface energy. It would be expected that the largest crystal faces would be those with the lowest surface energies. The fact that these surfaces were also those with the greatest interplanar spacings, d_{hkl}, corresponded to the law of Bravais (1866) [34], which stated that the larger the interplanar spacing, the more morphologically important (i.e. larger) the crystal face (*hkl*). In 1907 Friedel [35] provided convincing experimental examples to support this law, although violations were observed. Donnay and Harker [36] explained some of these anomalies by making allowances for crystal symmetry as implied by the space group: translational elements can cause the surface structure to be repeated more than once in the period d_{hkl}.

The Bravais–Friedel–Donnay–Harker (BFDH) law can be illustrated using the example of benzophenone ($(C_6H_5)_2C{=}O$) which crystallises in the orthorhombic space group $P2_12_12_1$. Each 2_1 screw axis lowers the morphological importance of the three principal axes, changing the morphological importance sequence from (010), (100), (001), (110), (111) to (110), (111), (020), (200), (002).

Using the BFDH principle and given the unit cell dimensions and space group, it is possible to rank the order of the morphological importance for a range of crystal habit planes, and computer programs for defining these faces are available [37]. However, this approach makes use of the framework of the crystal lattice but no consideration is made of the chemical bonding, which must have an effect during crystal growth.

It was Hartman and Perdok [38] who argued that, as the growth process was mediated through step motion on singular crystal surfaces, the use of only a surface-energy parameter to model the growth process was, perhaps, simplistic. Instead they preferred to use a new parameter, the surface attachment energy (E_{att}), defined [2] as the energy released on the addition of a growth slice to the surface of a growing crystal (Figure 7.4) to model the growth rate of the crystal surface. This term is related to the crystallisation (or lattice) energy (E_{cr}), i.e. the energy required to dissociate a crystal into its constituent molecules, by:

$$E_{\mathrm{cr}} = E_{\mathrm{sl}} + E_{\mathrm{att}} \tag{7}$$

where the slice energy E_{sl} is the energy released upon the formation of a slice of thickness d_{hkl}. The growth rate of a given crystal face is taken as proportional to E_{att} and so faces with the lowest attachment energies will be the slowest growing and have the most morphological importance [2].

This model assumes a number of factors, which can be rationalised by consideration of the growth of a flat crystal surface: the growth rate of a flat crystal surface is related to the enthalpy of dissolution at the growth interface ($\Delta H(T)_{\mathrm{diss}}$). If the dissolution of a crystallising unit fixed at a kink site on a 'flat' growth surface of a simple Kossel crystal is considered, then the molar enthalpy associated with bond (ϕ) formation at equilibrium can be derived [39]:

$$\Delta H(T)_{\mathrm{diss}} = N_{\mathrm{A}} \sum_{i=1}^{n} (\sigma_i^{\mathrm{of}} - 1/2(\phi_i^{\mathrm{ff}} + \phi_i^{\delta\delta})) \tag{8}$$

where N_{A} is Avogadro's number, the superscript $\delta\delta$ refers to solid–solid bonds at the growth interface and σf and ff refer to solid–fluid and fluid–fluid bonds in the bulk of the fluid. Due to the lack of structural data on solid–fluid interactions at the crystal-growth interface, and in order to use this theory on practical systems, three significant assumptions must be introduced:

(a) *equivalent wetting* – it is assumed that $\phi_i^{\sigma\mathrm{f}} = \phi_i^{\mathrm{sf}}$, i.e. that solid–fluid intermolecular bonds formed at the interface are the same as they would be if formed in the bulk solution, ϕ_i^{sf};
(b) *surface/bulk structure equivalence* – it is assumed that $\phi_i^{\delta\delta} = \phi_i^{\mathrm{ss}}$, where the superscript ss refers to solid–solid interactions in the bulk structure, and that no molecular rearrangement takes place at the surface;
(c) *the proportionality relationship* – it is assumed that the three bonds formed during crystallisation, ϕ_i^{ss}, ϕ_i^{sf} and ϕ_i^{ff}, are in the same ratio for any crystal surface (hkl), i.e.

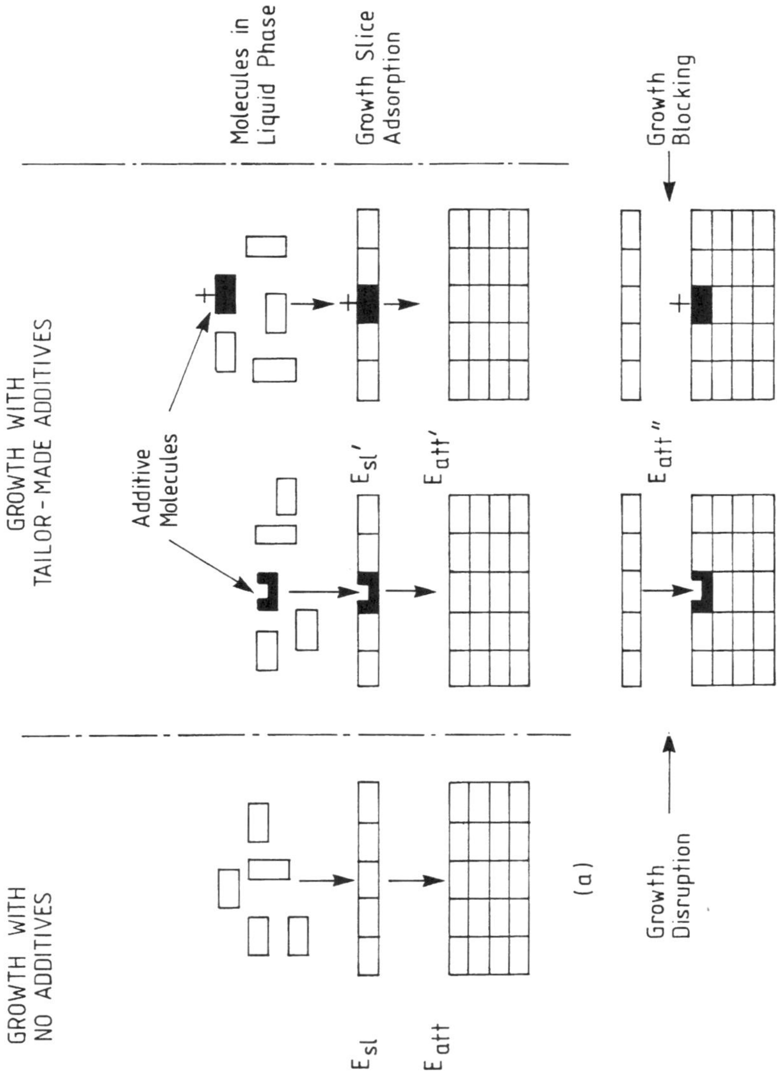

Figure 7.4 Schematic showing the definition of the energy terms used in morphological modelling for (a) pure systems and systems having (b) disruptive-type additives and (c) blocker-type additives

$$\phi_i^{ss}:\phi_j^{ss} = \phi_i^{sf}:\phi_j^{sf} = \phi_i^{ff}:\phi_j^{ff} \tag{9}$$

Using these assumptions, the growth process can be described in terms of only solid-state (ss) interactions.

For crystals growing at low supersaturations by either BCF or B&S mechanisms, Hartman and Bennema [2] have shown that:

$$R_{\text{growth}} \propto E_{\text{att}}^{ss} \tag{10}$$

Thus from the calculations of the intermolecular bonds (ϕ_i^{ss}) in relation to a given crystal surface (*hkl*), it is possible to model the relative growth rates of the crystal faces prominent in the external crystal morphology. Despite these assumptions, an extensive series of studies [5, 8, 9, 40, 41] have revealed a surprisingly reliable correlation between observed and predicted morphologies for molecular crystals. This probably reflects the fact that the nature of interactions between the molecules, in both the mother phase and in the resultant solid state following growth, is effectively dominated by relatively weak van der Waals forces.

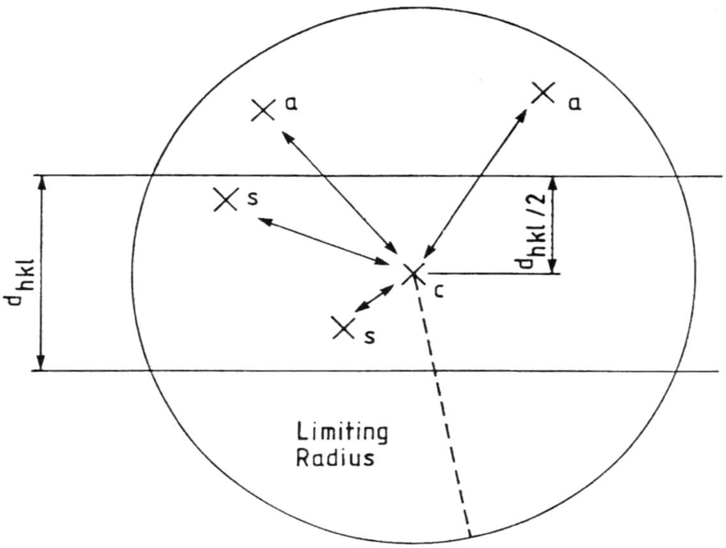

Figure 7.5 Basic approach for calculating the intermolecular interactions using the atom–atom method, showing how the lattice energy can be partitioned between the slice and attachment energies within a limiting sphere: X_c is the central molecule, X_a is a molecule outside the slice and X_s is a molecule inside the slice. Note that the slice boundaries defined by d_{hkl} may be shifted along the growth normal to obtain the most energetically stable slice

The overall methodology for morphological prediction can be summarised:

(a) Using the geometrical BFDH approach we can select the surfaces with large interplanar distances and hence high slice energies; from Eq. (7), these are expected to have low attachment energies and hence be the morphologically dominant faces of the crystal.
(b) The slice and attachment energies can be calculated by summing, pairwise, the potential energy involved in the intermolecular interactions between a central molecule and all the molecules within and outwith, respectively, a slice of thickness d_{hkl} (Figure 7.5). From this we can apply the 'atom–atom method' to evaluate the individual intermolecular bonds.

It is also necessary to consider the optimum position of the growth slice with respect to the d_{hkl} spacing. Figure 7.6 highlights the importance of this selection with the example of ε-caprolactam slice energies as a function of d_{hkl} spacing [24]. Here we see that while the optimum slice position for the {200}, $\{11\bar{1}\}$ and

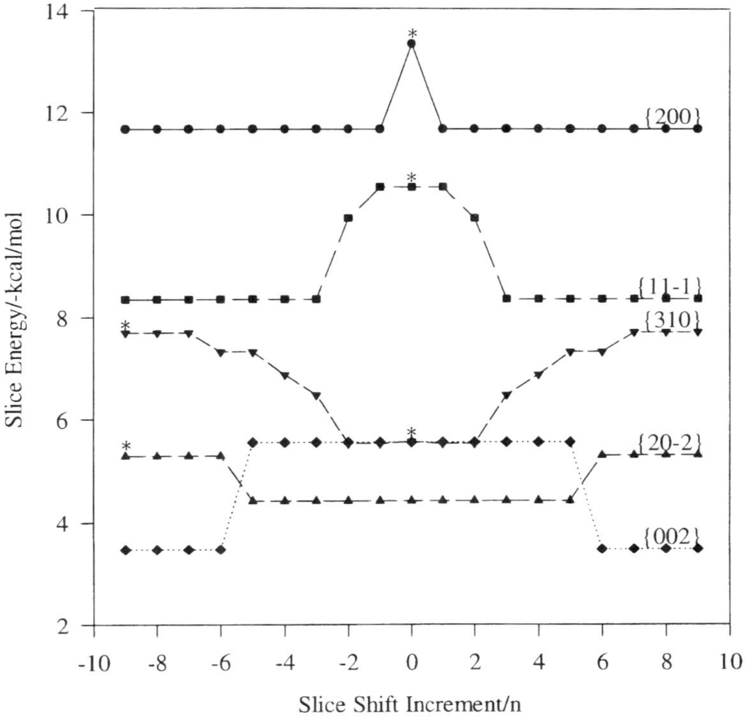

Figure 7.6 Demonstration of slice energy optimisation in the case of ε-caprolactam. Graph of slice energy (−kcal/mol) versus multiple of d_{hkl} spacing after [24]

$\{002\}$ forms is at the centre of the slice, when the $\{310\}$ and $\{20\bar{2}\}$ faces are considered, the position a distance $d_{hkl}/9$ from the centre gives the most stable slice. Thus the positioning of the slice with respect to the d_{hkl} spacing has a considerable impact on the resultant energies.

4.1 Examples [24, 41, 42]

Advances in the computational power of computer workstations mean that morphology prediction calculations can now be done in a fairly routine manner. These have been used to predict the crystal morphology of a wide range of organic molecular solids [10, 13, 24, 42, 43]. Some examples of crystal morphology predictions are shown in Figure 7.7a–f to highlight the flexibility of the method. These examples cover a wide range of industrially important materials, from organic explosives (pentaerythritol tetranitrate) to pharmaceuticals (nortriptyline hydrochloride) and materials of unusually high symmetry (cubic hexamine), all of which exhibit a range of morphologies from needles (terephthalic acid) to rhomboids (naphthalene) and plates (stearic acid). Table 7.4 gives the morphologically important faces produced from BFDH analysis, the d_{hkl} spacings and resultant attachment energies in the stearic acid system [20]. This highlights the reciprocal relationship between the two variables, and, when compared with the computed morphology, it can be seen that the face with

Table 7.4 Interplanar spacings and attachment energies for stearic acid

Growth form $\{hkl\}$	d_{hkl} (Å)	E_{att} (kcal/mol)
$\{001\}$	43.86	−1.40
$\{1\bar{1}\bar{1}\}$	4.26	−21.93
$\{1\bar{1}0\}$	4.12	−21.85
$\{1\bar{1}1\}$	3.97	−24.47
$\{020\}$	3.69	−29.71
$\{02\bar{1}\}$	3.68	−29.87
$\{20\bar{1}\}$	2.55	−34.02
$\{200\}$	2.48	−34.20
$\{201\}$	2.42	−35.35

Figure 7.7 Examples of morphological predictions: (a) pentaerythritol tetranitrate ($C_5H_8N_4O_2$), an organic explosive; (b) stearic acid ($CH_3(CH_2)_{16}CO_2H$), a long-chain carboxylic acid, containing hydrogen bonds; (c) naphthalene ($C_{10}H_{10}$), a molecule containing fused aromatic rings; (d) hexamine ($(CH_2)_6N_4$), an amine with unusually high symmetry (cubic space group); (e) nortriptyline hydrochloride ($C_{19}H_{22}HCl$), a pharmaceutically important organic salt; (f) α-terephthalic acid ($C_8H_6O_4$), an aromatic acid important in polyester production

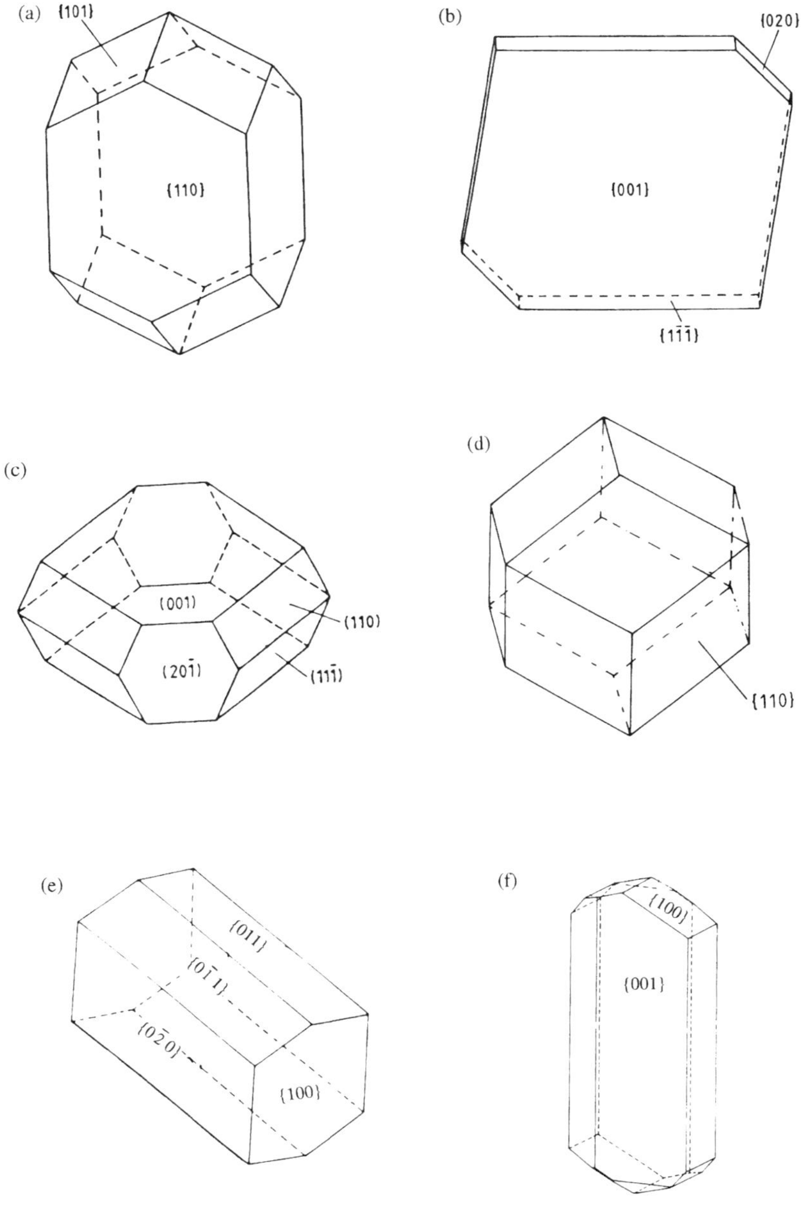
(a)
{101}
{110}
(b)
{020}
{001}
{1$\bar{1}\bar{1}$}
(c)
(001)
(110)
(20$\bar{1}$)
(11$\bar{1}$)
(d)
{110}
(e)
{011}
{0$\bar{1}$1}
{0$\bar{2}$0}
{100}
(f)
{100}
{001}

Table 7.5 Predicted aspect ratios of triclinic and monoclinic *n*-alkanes in the range $C_{18}H_{38}$ to $C_{28}H_{58}$

n-Alkane	Aspect ratio	
	Triclinic	Monoclinic
$C_{18}H_{38}$	7.1	8.1
$C_{20}H_{42}$	7.6	9.1
$C_{22}H_{46}$	8.9	10.2
$C_{24}H_{50}$	9.3	11.0
$C_{26}H_{54}$	10.3	12.0
$C_{28}H_{58}$	11.5	13.8

the largest d_{hkl} spacing and lowest attachment energy has by far the largest morphological importance. Table 7.5 also gives a list of predicted aspect ratios (a measure of the difference between the length and width of a crystal) from a series of triclinic and monoclinic *n*-alkane morphology predictions [41], indicating that as the chain length of the alkane increases, so the aspect ratio decreases.

5. EFFECT OF HABIT-MODIFYING ADDITIVES

It is well known that the growth of crystals is strongly influenced by the nature of the crystallising environment. Adding small amounts of impurities (additives) or changing the solvent or temperature often produces large changes to the growth rates of individual habit faces and hence can modify the morphology of the crystallised material. This process is essentially a case of 'molecular trickery' in which the hetero-molecule mimics, to a degree, the surface chemistry of the host system, thus 'fooling' the crystal surface into accepting it as a pure host molecule. Once in the surface, the impurity prevents further molecules getting to their rightful sites on the surface and alters the intermolecular bonding. Additive molecules have most effect when they interrupt a chain of strong bonding, e.g. hydrogen bonds [6].

It has been shown above how the overall crystal morphology can be related to the interaction energies between crystallising units. These energy calculations can be applied to determine the surface sites that can accept these solvent or additive molecules [6, 7] by calculating the relative 'binding' or 'incorporation energy' (Δb) of a crystal face as:

$$\Delta b = (E_{\mathrm{sl}'} + E_{\mathrm{att}'}) - (E_{\mathrm{sl}} + E_{\mathrm{att}}) \tag{11}$$

where $E_{\mathrm{sl}'}$ and $E_{\mathrm{att}'}$ are, respectively, the slice and attachment energy terms for the impurity-modified surface (see Figure 7.4). From this equation it can be seen that the crystal faces that have a minimum change in the binding energy (i.e. low values of Δb), are where the impurities are likely to incorporate. If Δb is strongly dependent upon crystal orientation then the incorporation will be specific to one crystal face and vice versa. In order to model the change of growth rate when an additive is present, the modified attachment energy parameter $E_{att''}$ is calculated. This reflects the energy released on the addition of a pure growth slice onto a surface on which an impurity molecule has adsorbed. $E_{att''}$ can be used as a direct measure of the growth rate of a crystal face 'poisoned' with an additive molecule, thus enabling calculation of the crystal morphology with an additive present. The modelling of tailor-made additives using this method has proved successful for a number of systems [8, 44].

In cases where the intermolecular interactions are unfavourable for steric reasons, vacancies can be introduced. This is achieved by calculating each intermolecular interaction and omitting its contribution to the energy terms, if it is considered to result from the blocking nature of the additive (for an illustration of this procedure see Figure 7.8). These steric repulsions are a result of atoms of the blocker residing close to, or actually in, the same physical space in the crystal as atoms of the adjacent host molecules. Effectively, removing an intermolecular interaction's contribution to the lattice energy is the same as omitting the molecule. This vacancy concept exhibits some of the main effects of blocker-type additives: a lowering in $E_{att''}$ (and hence an increase in morphological importance) and the loss of host molecules from adsorbing due to additive molecules blocking surface sites.

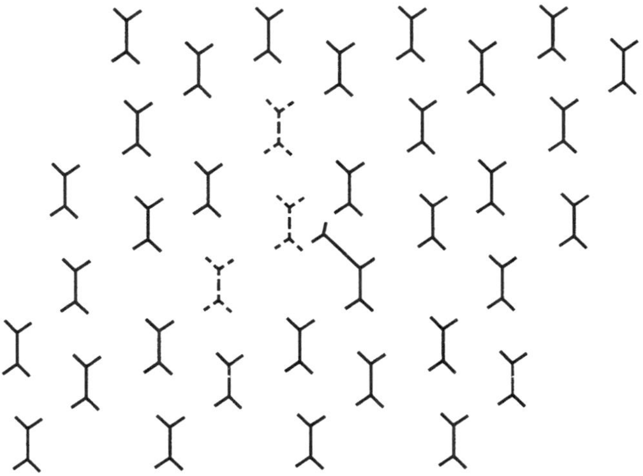

Figure 7.8 Example of the blocking nature of a 2-methyl-$C_{22}H_{46}$ additive molecule in a $C_{22}H_{46}$ crystal lattice, where for steric reasons, molecular vacancies (- - - - -) are created computationally)

5.1 Docosane ($C_{22}H_{46}$) [45]

Docosane crystallises in a triclinic space group with a plate-like morphology that results from strong interchain interactions and hence fast growth along the *a* and *b* directions. Thus large slow-growing (001) faces dominate the crystal morphology at the expense of the faster-growing (100) and (010) faces. The predicted morphology (Figure 7.9a) exhibits the expected plate-like habit dominated by the {001} form, agreeing with the observed crystal morphology [46].

The habit modification of docosane was modelled with the aim of predicting the effect of two disrupter-type additives with smaller chain lengths ($C_{20}H_{42}$ and $C_{21}H_{44}$) and two blocker-type additives with longer chain lengths ($C_{23}H_{48}$ and $C_{24}H_{50}$). The morphological impact of the *n*-alkane homologues can be seen from Figure 7.9b–e. These materials can incorporate easily into $C_{22}H_{46}$ crystals, although the habit-modifying properties are not great due to the similarity to the host system and the anisotropic nature of the bonding. The predicted crystals are thinner along the *c*-direction compared to the pure system, although they increase in thickness with chain length from 20 to 24. This is because of faster in-plane growth associated with the attachment energies of the side faces increasing due to more atoms being involved in the interchain interactions. These results reflect experimental observations, in that *n*-alkanes crystallised from solution [46] have, in general, much larger aspect ratios than those crystallised by sublimation [47].

The habit modification in the presence of simple mono- and dimethyl substituted *n*-alkanes (Figure 7.9(f–i)) is much more noticeable than was found with the *n*-alkane homologues. For example, the predicted crystal with 2-methyl docosane as an additive (Figure 7.9f) is much thicker along the *c*-axis, less wide along the *a*-direction and extended along the *b*-direction. The morphological impact is much greater than with $C_{23}H_{48}$ or $C_{24}H_{50}$ as an additive, i.e. the methyl group is a much more effective blocker. Table 7.6 lists the aspect ratios of the

Table 7.6 Predicted aspect ratios of pure $C_{22}H_{46}$ and additive-induced morphologies

System	Aspect ratio
Pure $C_{22}H_{46}$	12.5
$+C_{20}H_{42}$	17.9
$+C_{21}H_{44}$	15.4
$+C_{23}H_{48}$	15.6
$+C_{24}H_{50}$	14.0
+2-methyl-$C_{22}H_{46}$	9.7
+11-methyl-$C_{22}H_{46}$	9.7
+2,11-dimethyl-$C_{22}H_{46}$	8.5
+2,21-dimethyl-$C_{22}H_{46}$	7.9

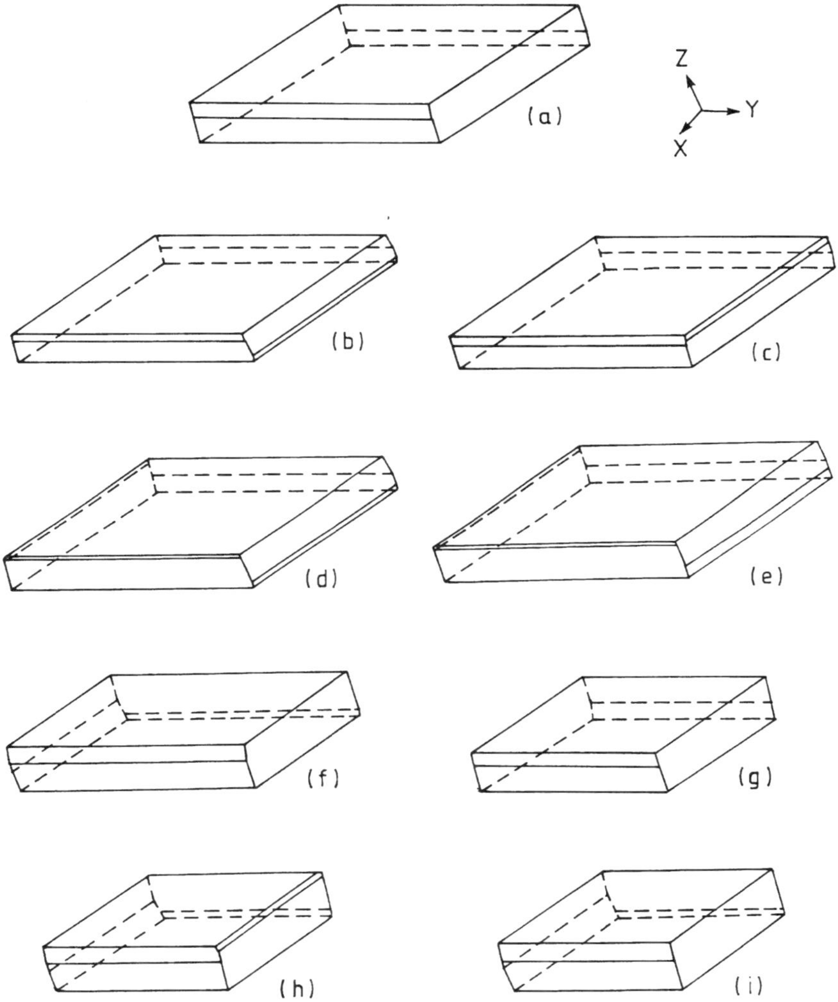

Figure 7.9 Simulated morphologies for $C_{22}H_{46}$: (a) pure $C_{22}h_{46}$; effect of the additives: (b) $C_{20}H_{42}$, (c) $C_{21}H_{44}$, (d) $C_{23}H_{48}$, (e) $C_{24}H_{50}$, (f) 2-methyl-$C_{22}H_{46}$, (g) 11-methyl-$C_{22}H_{46}$, (h) 2,11-methyl-$C_{22}H_{46}$, (i) 2,21-methyl-$C_{22}H_{46}$

host and additive systems and indicates that the addition of methyl groups to *n*-alkanes can have a significant effect on their morphology, decreasing the aspect ratio and forming more equant crystals. The effect is increased as a function of the number of methyl groups.

6. POLAR MORPHOLOGIES

Materials that crystallise in non-centrosymmetric space groups are likely to possess polar directions (i.e. crystals that exhibit different growth rates along (hkl) from $(\bar{h}\bar{k}\bar{l})$), reflecting unequal growth velocities for the crystal faces normal to the polar direction (this is the case for resorcinol [48]). Calculation of attachment energies between stoichiometric slices of crystalline materials (Figure 7.10) assume, by definition, a centre of symmetry in the calculation and thus $E_{att}(hkl)$ always equals $E_{att}(\bar{h}\bar{k}\bar{l})$. This problem, a manifestation of the three assumptions used in modelling theory (see Section 4), can be overcome by calculating the partial electronic charges on polar crystal surfaces and treating the polar growth process as one would treat an additive species. With this method, polar morphologies can be modelled by assigning a different set of atomic charges to molecules in the bulk of the crystal and to molecules on the surface of the crystal (these are the same as those for an isolated molecule). Such charges can be obtained using *ab initio* quantum-mechanical calculations via programs such as CRYSTAL [49].

6.1 Urea [50]

Urea crystallises in the highly symmetrical tetragonal space group $P\bar{4}2_1m$ with two molecules in a unit cell of dimensions $a = 5.576$ Å and $c = 4.684$ Å [51].

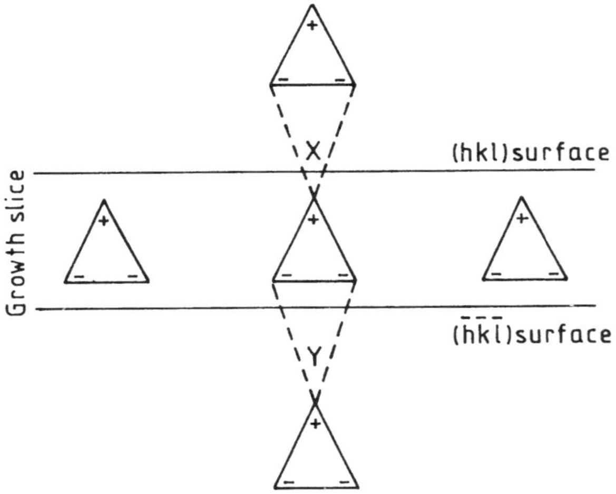

Figure 7.10 Schematic showing how the classical attachment energy model, based on the bulk solid-state structure, does not allow the prediction of polar effects; note the intermolecular bonds formed on attachment to the (hkl) face (X) are the same as those involved on the $(\bar{h}\bar{k}\bar{l})$ face (Y)

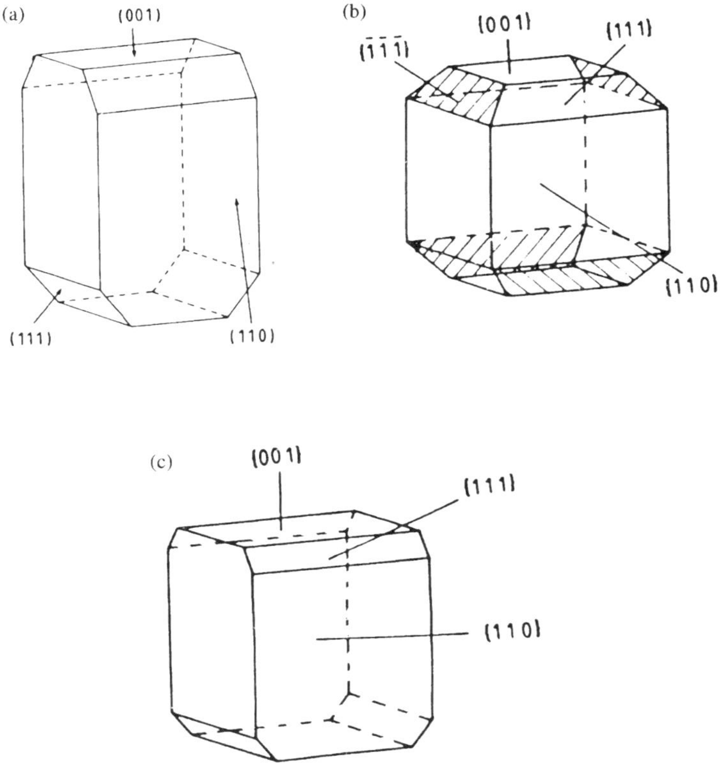

Figure 7.11 Urea morphologies: (a) experimental, (b) attachment energy prediction, (c) attachment-energy prediction using a polar morphology simulation

Crystals grown from the vapour phase have a well-defined prismatic morphology (aspect ratio 1.5) dominated by $\{110\}$, $\{001\}$ and $\{111\}$ forms, with the latter being polar, i.e. the $(\bar{1}\bar{1}\bar{1})$ face is not observed. Figure 7.11 illustrates (a) the experimental urea morphology along with the attachment energy predictions (b) normal and (c) in polar mode, demonstrating how polar attachment energy calculations can be used to predict such effects.

7. EFFECT OF GROWTH SOLVENT ON MORPHOLOGY [24]

Crystallising solvent species, much the same way as additives or impurities within a system, can cause changes in growth morphology. A variation in the growth solvent can also result in dramatic crystal habit alterations. A combination of Monte Carlo, molecular minimisation and molecular dynamics studies can be used to locate binding sites specific to each surface.

Solvent binding energies (E_{bind}) can be determined as the difference between the minimum energy of the sorbate (in this case, the solvent molecule) in contact with the surface and the energy of the sorbate (in its minimum-energy orientation) with no surface present (Eq. 12). By relating the binding energy of a sorbate molecule to that of a host molecule on the same surface, a differential interaction energy (E_{d}) can be determined (Eq. 13). These results are then used to scale the attachment energies using proportionality measurements (Eq. 14), i.e. for a value of E_{d} equal to -5 kcal/mol, the corresponding attachment energy was reduced by 10% (an arbitrary value). When E_{d} is positive, i.e. reflecting energetically unlikely solvent adsorption, the attachment energies are not altered. In this way a solvent-modified crystal morphology can be predicted.

$$E_{\mathrm{bind}} = E_{\mathrm{complex}} - E_{\mathrm{sorbate}} \tag{12}$$

$$E_{\mathrm{d}} = E_{\mathrm{bind}}(\mathrm{sorbate}) - E_{\mathrm{bind}}(\mathrm{host}) \tag{13}$$

$$E_{\mathrm{att}}(\mathrm{modified}) = E_{\mathrm{att}} - \left[\left(\frac{E_{\mathrm{d}}}{5} \times 10\right) \Big/ 100 \times E_{\mathrm{att}}\right] \tag{14}$$

7.1 ε-Caprolactam [52, 53]

ε-Caprolactam crystallises in a monoclinic structure [54] (space group $C2/c$; unit cell parameters $a = 19.28$ Å, $b = 7.78$ Å, $c = 9.57$ Å, $\beta = 112.39°$; eight molecules in the unit cell) in a solid-state structure dominated by dimer conformations associated with hydrogen bonds between carboxyl and amino groups of adjacent molecules. The predicted morphology was produced, using charge-equilibration [55] (QEq) atomic charges and is shown in Figure 7.12a, while the morphology of a crystal grown by sublimation is in Figure 7.12b. When grown from the vapour, ε-caprolactam crystals have a platy morphology with dominant $\{200\}$ forms, bounded by $\{110\}$ and $\{11\bar{1}\}$ faces. The prediction compares favourably with experimental evidence, the main difference being an overestimation in the growth rate of the $\{200\}$ face, resulting in the prediction of a thicker platelet.

Surfaces of interest – (200), (110) and $(11\bar{1})$ – which appear in the morphology, and $(31\bar{1})$ which is known to appear in the presence of certain growth solvents [56], were examined [57]. It was noted that both the (110) and $(31\bar{1})$ surfaces were rough with large voids present, which may encourage solvent incorporation. Also, the carbonyl and amino groups were exposed and therefore prone to attack from any oncoming solvent molecules with the ability to hydrogen-bond. The effects of the solvents acetone and ethanol were investigated on these surfaces. The method produced new solvent-modified attachment

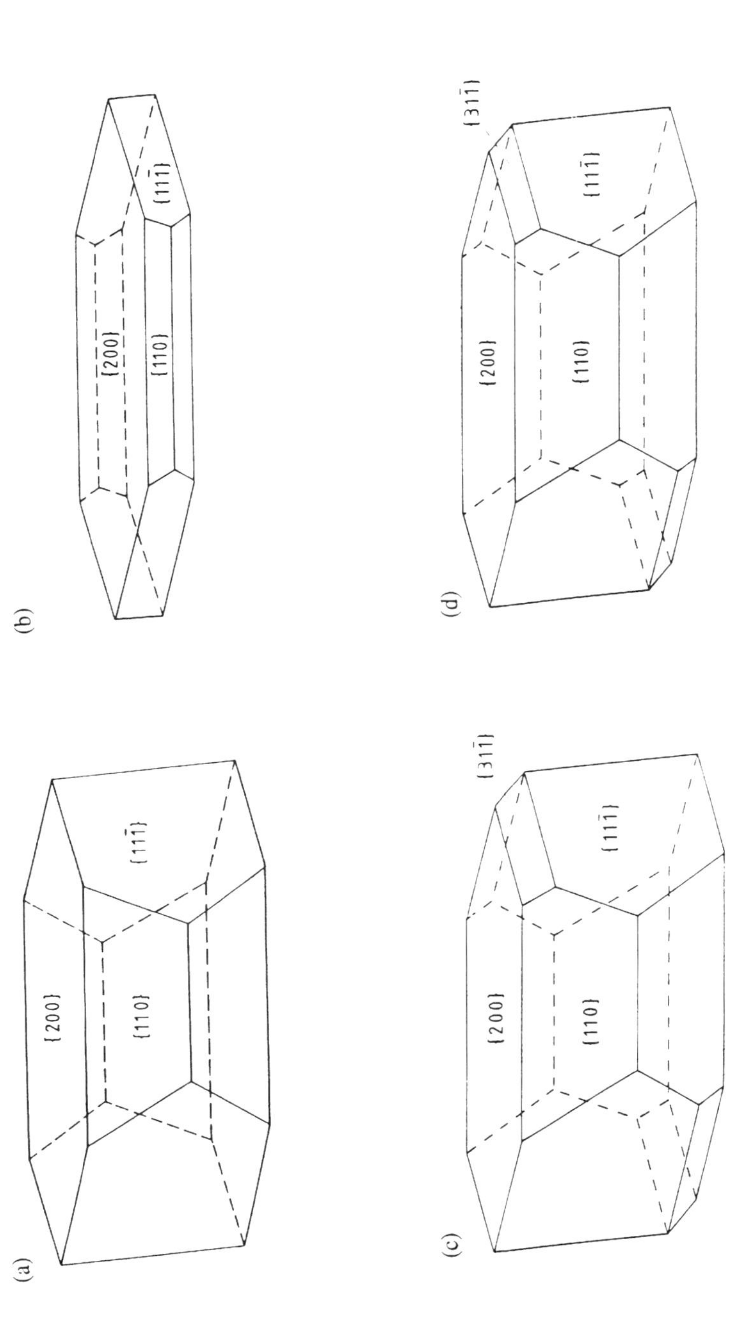

Figure 7.12 ε-Caprolactam: (a) predicted attachment energy morphology, (b) morphology from sublimation, (c) and (d) predicted solvent-induced morphologies form acetone and ethanol respectively

Table 7.7 Binding, differential interaction and attachment energies for ε-caprolactam solvent-modified system

	E_{att}	E_{bind}	E_d	E_{att}(modified)
Acetone				
{200}	−4.31	−7.24	+0.36	−4.31
{110}	−5.65	−14.71	−3.93	−5.20
$\{11\bar{1}\}$	−7.12	−8.18	+0.28	−7.12
$\{31\bar{1}\}$	−8.68	−12.34	−7.11	−7.46
Ethanol				
{200}	−4.31	−7.02	+0.57	−4.31
{110}	−5.65	−16.48	−5.69	−5.03
$\{11\bar{1}\}$	−7.12	−14.42	5.96	−6.27
$\{31\bar{1}\}$	−8.68	−16.72	−11.48	−6.86

energies (see Table 7.7) resulting in the appearance of the $\{31\bar{1}\}$ form for both solvents (see Figure 7.12c and d).

8. CHANGING MOLECULAR CONFORMATION DURING GROWTH

As described in Section 4, a number of assumptions are made in morphological predictions. However, when the growth process involves a change in molecular conformation, the adsorbing and host molecular species are likely to differ [58, 59]; therefore, the surface structure will not always be equal to that of the bulk crystallographic structure, implying a face (*hkl*) specific growth mechanism. That is, as surface chemistry changes with crystal orientation, molecular conformational change is likely to be specific to a particular crystal habit plane which will, in turn, change the surface-attachment energy. Such a change needs to be considered when trying to model the morphology of a material where a degree of molecular conformational flexibility can be expected.

8.1 Benzophenone [58]

Benzophenone crystallises in an orthorhombic structure with space group $P2_12_12_1$ and cell dimensions $a = 10.26$ Å, $b = 12.09$ Å and $c = 7.88$ Å [60] in a tetramolecular unit cell. The predicted attachment energy morphology is illustrated in Figure 7.13a alongside the experimentally produced shape in Figure 7.13b. The dominant differences between the observed and theoretically modelled morphologies is associated with the prediction of a more squat morphology than that experimentally observed. The normalised morphological data given in Table 7.8 shows that the agreement between experiment and theory for the ($hk0$)

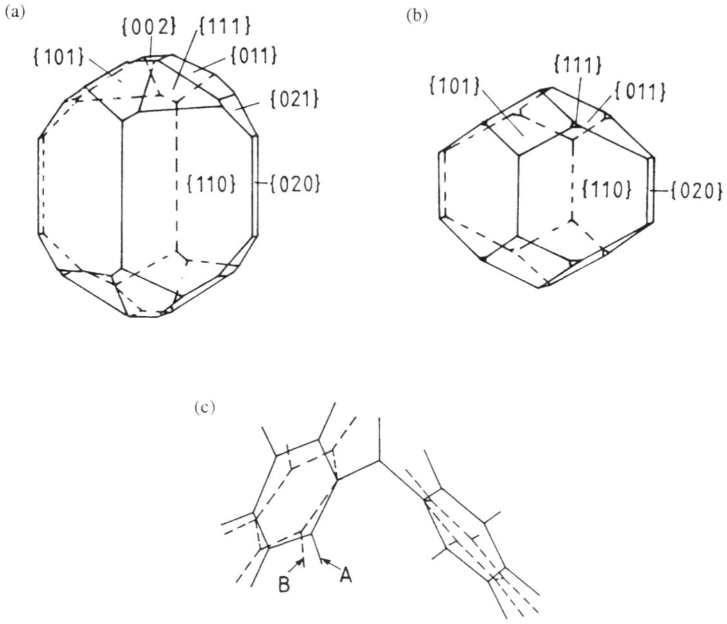

Figure 7.13 Benzophenone morphologies: (a) experimental, (b) attachment energy prediction, (c): conformational change of molecular structure (A) from crystal to isolated molecule (B)

and (001) forms is excellent and that the poor overall simulation reflects the fact that the attachment energy model overestimates the importance of the pyramidic {101} and {011} faces, resulting in the absence of the {021} and {002} forms.

A comparison of the molecular structure of benzophenone within the crystal and an isolated molecule revealed that a degree of conformational change was

Table 7.8 Normalised attachment energies for benzophenone against experimental data after scaling with respect to the {110} form

(*hkl*)	Calculated	Experimental	Difference (%)
(110)	1.00	1.00	0.0
(101)	1.25	1.56	24.8
(011)	1.17	1.47	25.6
(020)	1.42	1.43	0.7
(111)	1.37	1.47	7.3
(021)	1.45	1.50	3.4
(002)	1.66	1.67	0.6

taking place about the carbonyl group. Figure 7.13c shows these differences, noting that this conformational change can result in alterations in the crystal packing of the molecule on particular habit faces.

Effects due to the conformational change do not appear to be significant for adsorption on the $\{hk0\}$ crystal surfaces; however, on $\{0kl\}$ surfaces there is a greater degree of close packing, so that the number of significant atom–atom interactions can be expected to be much higher. As a result, calculations of surface-attachment energies based only on the solid-state structure, neglecting this conformational adaptation, seem to lead to an overestimation of the relative importance of the $\{hkl\}$ crystal forms, resulting in a smaller aspect ratio.

In terms of crystal growth theory this rationalisation demonstrates that the conformational-change enthalpy associated with the crystallisation process is highly face-specific. Thus, in the case of benzophenone (and benzil [58]) the proportionality approximation (see Section 4) appears to break down. This suggests that caution is needed when predicting the morphology of molecular crystals solely on the basis of the solid-state crystal structure when conformational flexibility is involved.

9. CLOSURE AND FORWARD LOOK

Due to the industrial importance of crystallisation and precipitation processes, the ability to model the external crystal shape of both homogeneous and heterogeneous systems is becoming increasingly important. Thus, methods that have evolved during the past century are currently finding wide and varied applications in both the fundamental and applied aspects of crystal science and engineering. These developments have been accelerated by advances in computational chemistry techniques that have led to the now routine use of morphological modelling approaches in improving our understanding and definition of the formation processes of organic solid particulates. The program HABIT95 allows morphological studies of pure systems, as well as assessing the effect of the crystal-growth environmental factors [8, 9, 24, 40, 41, 45, 52, 59, 61, 62]. The following extensions to the methodology are currently under development:

(a) the extension of these techniques to ionic molecular solids;
(b) modifications to account for blocker additives that contain hydrogen bonds;
(c) the use of surface modelling approaches to predict the structural nature of the growth interface when mediated by liquid phases;
(d) improved consideration of solvent effects through the study of surface relaxation, to obtain the optimum host surface structure;
(e) prediction of segregation coefficients to allow prediction of purification efficiency;
(f) extension of the overall methodology beyond the B&S growth mechanism

towards the consideration of surface defects inherent in the BCF screw-dislocation mechanism and rough-surface modelling needed for the RIG mechanism.

Current work and these resulting developments are all directed towards our Centre's strategic aim, which is to use a molecular-scale approach to understand, predict, control, manipulate and optimise the manufacture of solid chemical products.

10. APPENDIX: HABIT95 – A PROGRAM FOR PREDICTING THE MORPHOLOGY OF MOLECULAR CRYSTALS AS A FUNCTION OF THE GROWTH ENVIRONMENT

HABIT95 is a computer program designed to aid in morphological investigations on molecular materials. It represents a major development of an earlier program HABIT [43]. HABIT95 [10] is capable of predicting the external crystal shape based on a knowledge of the internal structure and of assessing the role of blocking or tailor-made additives [45]. Structural factors important in morphology such as predicting the morphology of molecular complexes, modelling polar morphology [50], and modelling conformational changes during growth [58] can also be studied. In addition, there is an option to select the most stable surface growth slice and a number of routines have been added that improve the examination of the important intermolecular interactions within molecular materials, aiding the understanding of crystallisation, the effects of impurities/solvent and polymorphism as a molecular recognition phenomenon.

HABIT95 allows the intermolecular interactions to be identified in terms of crystallographic direction, partitioned on to individual atoms, broken down to particular atom–atom contributions and sorted in terms of either strength or contact distance. Analysis of the intermolecular bonds is helpful in identifying strong bonds in a crystal during a periodic bond chain PBC analysis [40], the atom–atom bonds are useful to discover the source of unusual steric interactions (e.g., close contacts). Comparison of intermolecular bond strengths between host molecules and those involved in host/additive interactions can also be made [8, 9].

Other important features include the provision to consider only intermolecular bonds that fall between an upper and a lower bond strength in the energy summation. This allows, for example, the removal of a dominant intermolecular bond (e.g. a hydrogen bond) to assess its importance to crystallisation [63]. In addition, there is an interface with the crystal-drawing program SHAPE [64] allowing the simulated morphologies to be viewed. For additive calculations, the program writes two SHAPE input files to allow a direct comparison of the predicted undoped and doped morphologies. A further capability allows the effect of the additive to be seen graphically via typical commercial molecular graphics systems, a data file in CSSR format can be written containing the atomic co-ordinates of an additive molecule in a host crystal. A visual representation of important interatomic contacts can also be obtained by selecting a single intermolecular bond.

11. LIST OF SYMBOLS

(hkl)	Miller indices of crystal growth face
$\{hkl\}$	Miller indices of symmetry-equivalent crystal growth faces
R_{growth}	Crystal growth rate
S	Supersaturation
E_{cr}	Lattice (or crystallisation) energy of crystal
n	Number of atoms in central molecule
n'	Number of atoms in N surrounding molecules
N	Number of surrounding molecules
V_{kij}	Interaction energy between atom i in the central molecule and atom j in the kth surrounding molecule
A, B, C	Constants in interatomic potentials
$q_i q_j$	Fractional charges on atoms i and j
r	Interatomic distance
V_{exp}	Experimental lattice energy
ΔH_{sub}	Sublimation enthalpy
d_{hkl}	Interplanar spacing for face (hkl), thus slice thickness of growth slice (hkl)
E_{att}	Attachment energy of growth slice (hkl)
E_{sl}	Slice energy of growth slice (hkl)
$\Delta H(T)_{diss}$	Molar enthalpy of dissolution
$\phi^{\sigma f}$	Energy of solid–fluid bonds in bulk of the fluid
ϕ^{ff}	Energy of fluid–fluid bonds in bulk of the fluid
$\phi^{\delta\delta}$	Energy of solid–solid bonds at growth interface
ϕ^{sf}	Energy of solid–fluid bonds at growth interface
ϕ^{ss}	Energy of solid–solid bonds in bulk structure
Δb	Incorporation (or binding) energy of additive molecule
$E_{sl'}$	Slice energy with additive at centre of growth slice (hkl)
$E_{att'}$	Attachment energy of growth slice (hkl) containing additive on to pure surface
$E_{att''}$	Attachment energy of pure growth slice (hkl) on to surface containing additive
E_{bind}	Binding energy of solvent molecule to molecular surface (hkl)
E_d	Differential interaction energy between solvent and host molecule on molecular surface (hkl)
U, V, W	Multiples of the unit cell along the crystal directions a, b and c

12. REFERENCES

1. P. Hartman in *Crystal Growth: An Introduction*, edited by P. Hartman, North-Holland, Amsterdam (1973).

2. P. Bennema and P. Hartman, *J. Cryst. Growth*, **49**, 145 (1980).
3. Z. Berkovitch-Yellin, *J. Amer. Chem. Soc.*, **107**, 8239 (1985).
4. S. N. Black, L. J. Williams, R. J. Davey, F. Moffat, D. McEwan, D. E. Sadler, R. Docherty and D. J. Williams, *J. Phys. Chem.*, **94**, 3223 (1990).
5. R. Docherty, G. Clydesdale, K. J. Roberts and P. Bennema, *J. Phys. D: Applied Physics*, **24**, 89 (1991).
6. Z. Berkovitch-Yellin, J. van Mil, L. Addadi, M. Idelson, M. Lahav and L. Leiserowitz, *J. Amer. Chem. Soc.*, **107**, 3111 (1985).
7. L. Addadi, Z. Berkovitch-Yellin, I. Weissbuch, M. Lahav and L. Leiserowitz, *Topics in Stereochemistry*, **16**, 1 (1986).
8. G. Clydesdale, K. J. Roberts and R. Docherty, *J. Crystal Growth*, **135**, 331 (1994).
9. G. Clydesdale, K. J. Roberts, R. Docherty and K. Lewtas, *J. Crystal Growth*, **141**, 443 (1994).
10. G. Clydesdale, K. J. Roberts and R. Docherty, HABIT95 to be published via the Quantum Chemistry Program Exchange Program No 670, QCPE **16**, 1 (1996).
11. W. K. Burton, N. Cabrera and F. C. Frank, *Phil. Trans. Roy. Soc.*, **243**, 299 (1951).
12. G. H. Gilmer and K. A. Jackson, in *Crystal Growth and Materials*, edited by E. Kaldis and H. J. Scheel, pp. 80–114, North-Holland, Amsterdam (1977).
13. K. A. Jackson, in *Liquid Metals and Solidification*, pp. 174–186, American Society for Metals, Cleveland, Ohio (1958).
14. A. Bondi, *J. Phys. Chem.*, **68**, 441 (1964).
15. F. H. Allen, J. E. Davies, O. J. Johnson, O. Kennard, C. F. Macrae, E. M. Mitchell, G. F. Mitchell, J. M. Smith and D. Watson, *J. Chem. Inf. Comput. Sci.*, **31**, 187 (1991).
16. R. A. Buckingham and J. Corner, *Proc. Royal Soc.*, **189**, 118 (1947).
17. S. Lifson, A. T. Hagler and P. Dauber, *J. Amer. Chem. Soc.*, **101**, 5111 (1979).
18. F. A. Momany, L. M. Carruthers, R. F. McGuire and H. A. Scheraga, *J. Phys. Chem.*, **78**, 1579 (1974).
19. A. I. Kitaigorodski, *Tetrahedron*, **14**, 230 (1961).
20. G. Némethy, M. S. Pottle and H. A. Scheraga, *J. Phys. Chem.*, **87**, 1883 (1983).
21. D. E. Williams, *J. Phys. Chem.*, **45**, 3370 (1966).
22. S. L. Mayo, B. D. Olafson and W. A. Goddard, *J. Phys. Chem.*, **94**, 8897 (1990).
23. M. J. S. Dewar and W. Theil, *J. Amer. Chem. Soc.*, **99**, 4899 (1979).
24. E. M. Walker, *Ph.D. Thesis*, University of Strathclyde, Glasgow (1996).
25. M. Bailey and C. J. Brown, *Acta Cryst.*, **22**, 387 (1967); **C40**, 1762 (1984).
26. C. A. Lucchesi and W. T. Lewis, *J. Chem. and Eng. Data*, **13**(3), 389 (1968).
27. R. J. Davey, S. J. Maginn, S. J. Andrews, A. M. Buckley, D. Cottier, P. Dempsey, J. E. Rout, D. R. Stanley and A. Taylor, *Nature*, **366**, 248 (1993).
28. R. J. Davey, S. J. Maginn, S. J. Andrews, S. N. Black, A. M. Buckley, D. Cottier, P. Dempsey, R. Plowman, J. E. Rout, D. R. Stanley and A. Taylor, *J. Chem. Soc. Faraday Trans.*, **90**(7), 1003 (1994).
29. R. J. Davey, S. J. Maginn, S. J. Andrews, S. N. Black, A. M. Buckley, D. Cottier, P. Dempsey, R. Plowman, J. E. Rout, D. R. Stanley and A. Taylor, *Mol. Cryst. Liq. Cryst.*, **242**, 79 (1994).
30. M. Saska and A. S. Myerson, *Cryst. Res. Technol.*, **20**(2), 201 (1985).
31. L. Leiserowitz, *Acta Cryst.*, **B32**, 755 (1976).
32. J. W. Gibbs, *Trans. Acad. Connecticut Acad.*, Vol. 3 (1875); see also *The Equilibrium of Heterogeneous Substances*, Scientific Papers Vol. 1 (1906) and *Collected Works*, Longmans Green, New York (1928).
33. G. Wulff, *Z. Krist.*, **34**, 499 (1901).
34. A. Bravais, *Études Crystallographiques*, Gauthier-Villars, Paris (1866).
35. G. Friedel, *Bull. Soc. Franc. Mineral.*, **30**, 326 (1907).

36. J. D. H. Donnay and D. Harker, *Amer. Mineralogist*, **22**, 446 (1937).
37. R. Docherty, K. J. Roberts and E. Dowty, *Comp. Phys. Comm.*, **51**, 423 (1988).
38. P. Hartman and W. G. Perdok, *Acta Cryst.*, **8**, 49 (1955).
39. P. Bennema and J. P. van der Eerden in *Morphology of Crystals*, pp. 1–75, edited by I. Sunagawa, Terra Scientific, Tokyo (1987).
40. G. Clydesdale, K. J. Roberts and R. Docherty in *Controlled Particle, Droplet and Bubble Formation*, pp. 119–121, edited by D. Wedlock, Butterworth-Heinemann, London (1994).
41. G. Clydesdale, PhD Thesis, University of Strathclyde (1991).
42. M. L. MacCalman and K. J. Roberts, *J. Crystal Growth*, **128**, 1218 (1993).
43. G. Clydesdale, K. J. Roberts and R. Docherty, *Comp. Phys. Comm.*, **64**, 331 (1991).
44. M. Lahav, L. Leiserowitz, Z. Berkovitch-Yellin, J. van Mil, L. Addadi and M. Idelson, *J. Amer. Chem. Soc.*, **107**, 3111 (1985).
45. G. Clydesdale, K. J. Roberts and K. Lewtas, *Mol. Cryst. Liq. Cryst.*, **248**, 243 (1994).
46. D. H. M. Beiny, J. W. Mullin and K. Lewtas, *J. Crystal Growth*, **102**, 801 (1990).
47. K. Lewtas, unpublished results (1993).
48. F. C. Wireko, L. J. Shimon, F. Frolow, Z. Berkovitch-Yellin, M. Lahav and L. Leiserowitz, *J. Phys. Chem.*, **91**, 472 (1987).
49. S. Dovesi, C. Pisani, C. Roetti, M. Causa and V. Saunders, CRYSTAL88, QCPE program number 577, Bloomington, Indiana, USA (1989).
50. R. Docherty, K. J. Roberts, R. J. Davey, S. N. Black and V. R. Saunders, *J. Chem. Soc. Faraday Discussions*, **95**, 11 (1993).
51. P. Vaughan and J. Donohue, *Acta Cryst.*, **5**, 530 (1952).
52. K. J. Roberts and E. M. Walker, *IChemE Research Event*, Institute of Chemical Engineers, 728 (1994).
53. K. J. Roberts and E. M. Walker, *Mol. Cryst. Liq. Cryst.*, **279**, 233 (1996).
54. F. K. Winkler and J. D. Dunitz, *Acta Cryst.*, **B29**, 268 (1975).
55. A. K. Rappé and W. A. Goddard III, *J. Phys. Chem.*, **95**, 58 (1991).
56. R. M. Geertman and A. E. D. M. van der Heijden, *J. Cryst. Growth*, **125**, 363 (1992).
57. *Cerius²* Molecular Modelling Software for Materials Research by Molecular Simulations Incorporated, Burlington, Mass. and Cambridge, UK.
58. K. J. Roberts, R. Docherty, P. Bennema and L. A. M. J. Jetten, *J. Phys D: Appl. Phys.*, **26**, B7 (1993).
59. G. P. Hastie, J. Johnstone, K. J. Roberts and E. M. Walker, *Mol. Cryst. Liq. Cryst.*, **277**, 29 (1996).
60. G. M. Lobanova and E. B. Vul, *Sov. Phys. Cryst.*, **12**, 355 (1967).
61. M. L. MacCalman, PhD Thesis, University of Strathclyde, Glasgow (1996).
62. R. Docherty, PhD Thesis, University of Strathclyde, Glasgow (1989).
63. K. J. Roberts, J. N. Sherwood, C. S. Yoon and R. Docherty, *Chem. Mater.*, **6**, 1099 (1994).
64. E. Dowty, *Amer. Mineralogist.*, **65**, 465 (1980).
65. J. D. Cox and G. Pilcher, *Thermochemistry of Organic and Organometallic Compounds*, Academic Press, New York (1970).
66. A. Aihara, *Bull. Chem. Soc. Jpn.*, **B2**, 1242 (1959).

Index

ff following
MD Molecular Dynamics
lit literature survey
tab table
fig figure

Additive-induced morphologies,
  docosane, 220
Additives, blocking, 219
  habit-modifying, 218
  tailor-made, 219
AMBER computer programs package, 120
Angular distribution function, 163
Aspect ratio, 218
Asymmetric unit, 2, 101
Atom–atom approach, 66
  force fields, analytical form, 96
  potential method, 91lit, 207
    parameters, 67
      combination rules, 53, 209
  potentials, 31–34, 121 *see also* Intermolecular potential
    anisotropic, 45
    for MD of clusters, 181
    transferability, 47
Atomic point charges, 36, 37, 188
Atomic polarizability, 132
Atoms in molecules, Bader's
  partitioning, 37
Attachment energy, benzophenone, 227tab
  ε-caprolactam, 226tab
  stearic acid, 216
Autocorrelation function, 118
Axilrod–Teller term, 41

Band theory, 88
Band, vibrational, 64
Basis-set superposition error
  (BSSE), 44, 92
Benzophenone, crystal morphology, 226ff
Birth-and-spread growth mechanism, 205
Bloch functions, 63
Born–Oppenheimer approximation, 99
Bravais theory, crystal growth, 211
Brillouin zone, 109
Buckingham potential, 66, 122, 208
Burton–Cabrera–Frank theory, 204
t-Butylchloride clusters, solid state
  transformations, 179

CADPAC computer programs package, 51
Cambridge Structural Database, 67, 87, 94,
  102, 189, 207
Capillary theory, homogeneous
  nucleation, 148
ε-Caprolactam, morphology
  prediction, 224
CCP5 program library, 182
Cell model, for lattice vibrations, 110
Cellulose, MD modeling, 135ff
Central multipole expansion, 41
Charge density, X-ray, 37
Charge group approximation, 127
Charge transfer, 43

CHARMm computer programs package, 120, 188
Chlorine crystal structure, 44
Classical trajectories in MD, 113
Close-packing, 22
and symmetry, 23
principle, 32
Cluster analysis, of predicted crystal structures, 193
Computer program packages: AMBER, 120; CADPAC, 51; CHARMm, 120, 188; CCP5, 182; CRYSTAL, 222; DMAREL, 50; GROMOS, 120; HABIT, 228–229; ICE9, 96; MACSPIN, 162; MDMPOL, 182; MM3, 120; MOLPAK, 95; MPA/MPG, 96; Polymorph Predictor, 188; PROMET, 27; SHAPE, 229
Conjugate gradient method, 111
Convergence acceleration, efficiency, 130
Coordination sphere patterns, 197
Core and surface energy in clusters, 178
CPU time, for MD of cluster phase transitions, 181
Critical nuclei, 167
Crystal disorder, 17, 65, 90lit, 104, 116, 134
dissolution, 82
engineering, 25
environment and intramolecular structure, 126
faces, growth rate modeling, 214
growth, 204
and molecular conformation, 226
theories, 211ff
habit, and solvent, 224
morphology, solvent-modified, 224
urea, 223
prediction, benzophenone, 226
ε-caprolactam, 224
docosane, 220
nucleation, 204
packing, 18
and molecular conformation, 228
prediction, 26
potentials, 65
structure prediction, 85, 95lit, 138ff
and space group symmetry, 189
computer approaches, 95
flexible molecules, 192
of azabenzenes, 49
of hexopyranose crystals, 140
CRYSTAL computer programs package, 222
Cutoff distance, 50, 114, 127, 128

Debye temperature, 109
Density of states, 63–64
p-Dichlorobenzene, polymorphs, 57
Dielectric constant, 124
Diffuse scattering, 102, 109
Dispersion energy, formula, 41
Dispersion (van der Waals) forces, 19
Displacement temperature factors, anisotropic, 133
Dissolution kinetics, 95lit
rate, 82
Distributed multipole analysis, Stone's, 37
model, computer program for, 49
performance, 51
DMAREL computer programs package, 50
Docosane, morphology, 220
Donnay–Harker theory, crystal growth, 211
Dynamical matrix, 64, 108, 128

Einstein crystal, 117
Elastic constants in crystals from MD, 116
properties of crystals, 89
Elasticity tensor, 102
Electric properties of crystals, 102
Electron correlation, 34, 38
exchange and attractive forces, 42
Electrostatic energy, 123
formula, 35
multipole expansion, 36
forces, importance of to crystal structures, 48
interactions, 207
Energy band, 63
Energy minimization methods, 53, 111
Enthalpy and crystal symmetry, 21
of combustion, 70
of dissolution, 78, 212
of guest desorption, 81
of melting, 70, 73, 75tab, 93lit
computer model, 73
of sublimation, 55fig, 67, 68tab, 70, 72tab, 92lit, 102, 209
estimation, 86, 97
Entropy and crystal symmetry, 21
changes during gas expansion, 153
of crystals, 104

from MD, 117
of melting, 75tab, 76
of sublimation, 71, 72tab
Equilibrium atomic positions, 101
of dissolution, 78
solid–liquid, 73
solid–solid, 77
solid–vapor, 70
Equivalent positions, 101
Equivalent wetting assumption, 212
Evaporative cooling temperature, 155, 167
Ewald sums, 50, 129, 130, 188
Excess thermodynamic functions, 79
Exchange repulsion, 42

Force field, consistent, 91, 119
definition, 99
general form, 120
parameterization, 119
sempirical, 91lit
survey, 120, 125
Forces on atoms, 101
Fractional coordinates, 101
Free energy barrier to nucleus formation, 170
of crystals, 108
differences, from MD, 117
of dissolution, 78
of melting, 73
Freezing of liquid clusters, MD, 178
Friedel theory, crystal growth, 211

Generalized momentum, 103
Glide plane, 7
GROMOS computer programs package, 120
Growth rate of crystal faces, 214

HABIT computer programs package, 228–229
Hamiltonian, in MD, 103
Hard sphere model, 126
Hartman–Perdok theory, crystal growth, 212
Heat capacity (specific heat), 70, 97, 102
Heat of *see* Enthalpy of
Hexafluorides, solid-state transformations, 180
Hickson–Crowell equation, 83
Hill potential, 122
Homogeneous nucleation, capillary theory, 148
Host–guest cocrystallization, 81
Hydrogen atoms coordinates, 62, 88, 133
bonding, 207
bond potentials, survey, 124

Ice, polymorphs, 176
Ice–water interfacial free energy, 175
ICE9 computer programs package, 96
Inclusion compounds, 80
Incorporation energy, 218
Induced dipoles, 131
Interfacial free energy, 158ff, 169, 171
and enthalpy of fusion, 172
ice–water, 173
liquid–solid, 174
Intermolecular potential, *see also* atom–atom potential
Barker, 34
from molecular overlap, 42
Inversion center, 4

Lattice, 2
Lattice dynamics, 89lit
energy, 68, 207 *see also* Packing energy
and enthalpy of sublimation, 97
atomic contributions to, 210
sums, convergence acceleration, 128
vibrations, 63, 107
cell model, 110
Lattice vibrational modes, 108
Leap-frog algorithm, 113
Legon-Miller rule, 46
Lennard–Jones potential, 121, 208
Lindemann index, 163ff

MACSPIN computer program, 162
MDMPOL computer program, 182
MPA/MPG computer programs package, 96
Melting temperature, 73
Metric matrix, 100
Metropolis algorithm, 192
MM3 computer programs package, 120
MNDO method, 209
Molecular dynamics, classical, 160
definition, 112
for crystal packing, MDCP method, 197
of benzene crystals, 197

Molecular dynamics, classical (*cont.*)
of monosaccharide crystals, 133ff
simulations, constant T and P, 114
trajectories, and uncertainty principle, 182
homeomorphism coefficients, 79
mechanics, 110ff
orientation match indexes, 166
reorientations, 24, 65, 90lit
MOLPAK computer programs package, 95
Monoclinic system, 7
Monosaccharide crystals, molecular dynamics, 133ff
Monte Carlo method, 118, 160
Simulated Annealing, 191
Smart, 190, 198
Morokuma decomposition, 43
Multipole expansion, of polarization energy, 39

Neutron scattering, inelastic, 109
Newton–Raphson method, 112
Newton's equations of motion, 112
Niggli reduced cell, 193
Non-bonded energy, derivatives, 127
forces on atoms, 127
interactions, calculation, 121
Coulomb, 123
1–4, 123
sums, 127–128
Noyes–Whitney equation, 83
Nozzle flow, 152
Nucleation, capillary theory, 172
and surface layers, 168
experiments, Turnbill's, 149
on cooling in clusters, MD, 178
rate, 159, 167–169
of clusters, measurement, 173
of solid-state transformations, 177
theory, 149ff
timescales, in MD, 151
Nucleus formation free energy barrier, 170

Off-ridge eigenvector minimization, 196–197
Optical properties of crystals, 89
Packing coefficient, 22
energy, 68, 70 (*see also* Lattice energy)
and solid solutions, 80
of one-dimensional chains, 195
of layers, 195
problem, 186
Pair correlation function, 163
Partition function, 103
Pauli's principle, 19, 42
Pawley projection, 162ff
Penetration energy, 45
Perturbation theory, 35
second order, 39–43
Phase space probability density, 104
transitions, 147
dynamics, 148
electron diffraction monitoring of, 150
in supercooled systems, 148
solid–solid, 118
solution-mediated, 84
temperature, effect of cluster size, 164
Plastic phases, 24, 77
Polar growth, 222
Polarizability, dipole–dipole, 39
distributed, Stone's analysis, 40
in force fields, 125
Polarization, 188
energy, 39, 131
model, 130
multipole expansion, 39
Polar morphology, 222
Polymorphism, 77, 78, 93–94lit, 104, 110, 186
Polymorph Predictor computer programs package, 188
Polymorphs, disappearing, 94
of benzene, 196
of ice, 176
of terephthalic acid, 209
Polypeptides, molecular charge distribution of, 48
Positional melting, 74, 93
Potential energy, equilibrium, 110
function, global minimum, 187
local minima, 191
Potentials, pairwise additive, *see* Atom–atom potentials
Powder diagrams, 194
Prediction of crystal morphology, 215ff
Preexponential factor, in nucleation rate, 169
Pressure tensor in crystals, 105, 106, 141
Programs and program packages, *see* Computer
PROMET computer programs package, 27